Geological Maps and Sections for Civil Engineers

GEOLOGICAL COLUMN

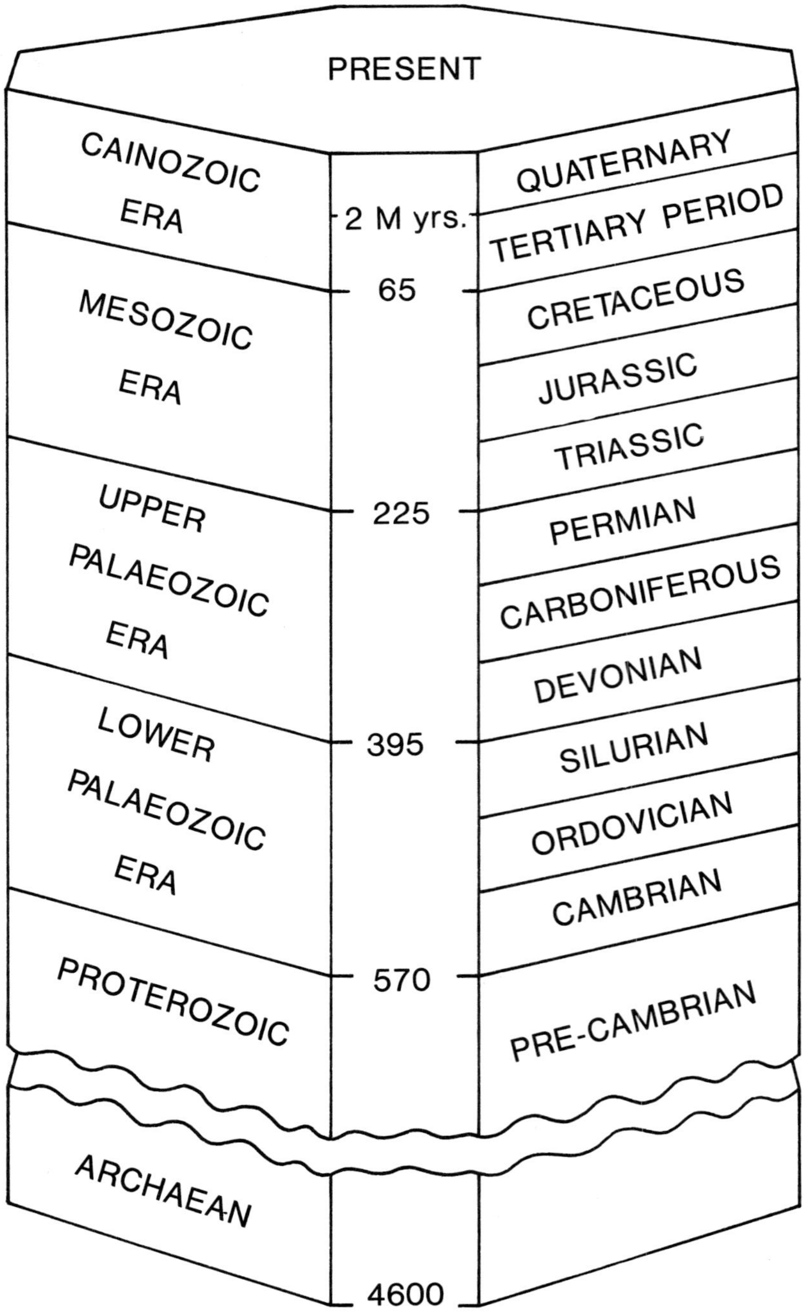

Geological Maps and Sections for Civil Engineers

PETER R. THOMAS
Department of Civil Engineering
University of Strathclyde

formerly of

Department of Civil Engineering
Paisley College of Technology

Blackie
Glasgow and London

Published in the USA and Canada by
CRC Press, Inc.
Boca Raton, Florida

Blackie and Son Ltd
Bishopbriggs, Glasgow G64 2NZ
and
7 Leicester Place, London WC2H 7BP

Published in the USA and Canada by
CRC Press, Inc.
2000 Corporate Blvd, N.W., Boca Raton, FL 33431

First published 1991

British Library Cataloguing in Publication Data

Thomas, Peter R.
Geological maps and sections for civil engineers.
1. Geological maps. Map reading
I. Title
551.8

ISBN 0-216-92903-2

Library of Congress Cataloging-in-Publication Data

Geological maps and sections for engineers / P. R. Thomas.
p. cm.
Includes bibliographical references.
ISBN 0-8493-7142-2
1. Geology—Maps. 2. Engineering geology—Maps. I. Thomas, P. R.
QE36.G44 1990
550′.22′3—dc20 89-22075
CIP

Preface

In writing this book the author's aim has been to assist communication and understanding between civil engineers and geologists when dealing with maps and sections. It is written specifically with civil engineering students in mind, so that they may understand the fundamentals of geological map reading and become more aware of the value of geotechnical mapping techniques. It will also help geologists to interpret traditional geological maps and to learn about specific techniques which provide data of particular use in civil and geotechnical engineering. The book has been designed to tackle problems logically, from a basic appreciation of the use and geometry of maps, through to the techniques which are likely to be used by the geologist and the practising engineer when communicating about ground conditions.

Civil engineers may have little appreciation of the usefulness and limitations of geological maps as they relate to civil engineering. Yet maps and plans are an important part of most civil engineers' daily work programmes. Civil engineering students may take readily to the geometrical relationships of the geological map and the high degree of accuracy of man-made structural details, but fail to appreciate that such precision is not easily obtainable from complex, natural phenomena. Good engineers learn to adjust to the variable conditions which are encountered when working with the products of earth processes.

This book will enable the civil engineer and engineering student to appreciate fully the aids provided by geologists to the interpretation of ground conditions. At the same time, it will give the geologist valuable insight into the requirements of practising civil engineers faced with either an initial appraisal of geological conditions at a proposed site, or unexpected conditions which may have arisen during the construction process.

Texts on ground investigation do exist, but this book deals primarily with the *early planning* and *exploration* aspects of such investigations. It provides guidance on using and understanding geological maps as sources of information for specifying a site investigation borehole programme. Although the book is aimed mainly at students, it will be of value to practising engineers and geologists requiring an update on the use and availability of geological maps, and on the developing role of specialised geotechnical maps based on intensive local investigations.

Acknowledgements

I would like to thank all those in the Paisley College Civil Engineering Department who helped in various ways during the writing of this book. Special thanks go to Professors J. F. Woodward and W. B. Cranston for their encouragement during the last three years. Dr Judith Lawson gave advice on the initial manuscript, which was typed by Linda McDermid. Practical help was also willingly given by technicians Yvonne Fraser and David Wallace. The constructive criticism of Ian Higginbottom helped me enormously, as did the expert guidance and optimism of the publishers.

I acknowledge the British Geological Survey for permission to reproduce maps and diagrams, and Professor Bill Dearman, whose work has been an inspiration.

PRT

To my wife, Helen

Contents

Chapter 1 The geological map

1.1 Introduction

Throughout the world geological surveys have been, and are being, carried out by official agencies in order to produce a comprehensive record of the distribution of rocks and superficial deposits at the surface.

Superficial deposits are produced by relatively recent sedimentary processes at the earth's surface and are rarely sufficiently cemented or consolidated to be considered as rock for excavation purposes. Engineers regard most superficial materials as *soils*, although some weakly cemented rocks may also be classed by engineers as soils. Different disciplines tend to adopt differing terminologies, particularly with regard to the definition of soil. For example, academic geologists, geographers and soil scientists use the term in an agricultural sense and the term *subsoil* or *Drift* is used in connection with superficial deposits underlying the first metre or so of the soil profile.

In the early history of geology the name Drift came to be applied as a convenient shorthand for superficial deposits and the underlying bedrock represented the *Solid* base on which the Drift deposits were laid down. Hence the adoption of two different styles of geological map named Drift and Solid editions arose.

Drift maps show engineering soils and areas of outcropping or shallow rock. Solid editions of geological maps predict the pattern of underlying strata swept clean of any superficial deposits. The two editions covering the same area are commonly made available in Britain so that both clarity and continuity of rock pattern may be achieved. This practice may be peculiar to North Britain where blanket drift cover is not uncommon and yet occasional exposures, boreholes and underground workings for coal and other minerals provide enough details to enable a Solid edition to be published. In unglaciated areas, river deposits, weathering and periglacial deposits may be restricted enough to allow a combined *Solid and Drift* map to be produced. This is similar to the Drift editions but does not attempt to indicate the solid rocks below the drift. In fact many geological maps throughout the world are of this type, virtually providing information about what is found at the surface and not attempting to predict rock in poorly exposed areas covered by drift. It should be noted that *Soil Survey* maps, actually depicting pedological types, are also available and may be useful in detecting subtle changes in subsoil conditions.

1.2 The engineer's viewpoint

The basic types of geological map, described above are traditional in Britain. They are not produced with any one group of users in mind and older maps may be based on surface observations using little more than an auger to assess the drift cover. Classification of materials is usually geological rather than geotechnical, so that engineers can only appreciate their value if they have some knowledge of how the maps were made and what limitations have to be imposed on their use when tendering for or planning civil engineering operations. Unless the engineer is prepared to pay for a new geological survey as part of a feasibility study, he normally has to rely on those maps already available either in published or draft form.

Most British Geological Survey maps are prepared initially as field slips at a scale of 1:10 560, or more recently 1:10 000, but in large countries this scale is impractical and smaller scales of 1:50 000 to 1:250 000 are necessary. Even the scale of 1:10 000 used in Britain is not always adequate for showing detail in complex areas and it is scale which often creates the greatest problem for engineers, particularly when many site plans have scales between 1:2500 and 1:100.

It is not unknown for an engineer to transfer a

geological boundary from a 1:50 000 map on to a 1:1250 site plan and subsequently complain that the boundary was found to be 50 m out of place on the site! But it is not only the scale which limits the precision of boundaries. For example, since it is not normal practice to publish solid geological maps with blank areas (except where lakes or sea areas are involved), the poorly exposed parts of the maps may sometimes be a best estimate of the position of boundaries. It is only with the prudent combined use of both the Solid and Drift sheets that the degree of accuracy of the solid geology can be assessed.

1.3 Drift maps

These geological maps are particularly useful for preliminary reference or route planning prior to site investigation. They show more recent surface deposits including peat, river alluvium, marine and estuarine sediments, most of which overlie glacial or periglacial deposits in Britain. Only the larger landslip areas are plotted. They also give the engineer an indication of where rock outcrops over large areas or is very near to the surface. Drift maps produced by the Geological Survey in England and Wales display the various rock types which outcrop, whereas more recent maps in Scotland and some in northern England show all rock areas in grey, necessitating the combined use of both Solid and Drift editions. It is useful at the planning stage to have some idea of the soil–rock distribution on an extensive site, but apart from the scale problem mentioned above, the *classification* of drift may be much too broad compared with classifications used by geotechnical engineers. Specific data on *depth* of drift is lacking on small scale maps and it is only by intelligent consideration of outcrop distributions that any estimate of depth can be attempted. The same may be said about *drift successions*, since only the uppermost 'soil' is shown. Underlying layers can only be inferred from a careful study of the distribution and interrelationship of drift across the map and perhaps some knowledge of the recent geological history of the area—a factor sometimes left out of geotechnical studies.

Modern Drift maps are enhanced by the feedback of information from previous site investigations, which may include both boreholes and geophysical surveys. The *boundaries* are usually more accurate on Drift maps since surface features and auger holes or pits allow lines to be drawn with more confidence, especially in rural areas. Air-photo interpretation and remote sensing are improving accuracy in most recent work, but in the urban situation and especially industrial wastelands, natural landforms are becoming increasingly disturbed by man's activities. The distribution of landfill and made-ground are generally shown only on the most recent Drift maps. In this respect, old topographical maps prove invaluable when tracing the former uses of what may appear to be greenfield sites.

1.4 Solid geological maps

Where exposure of rock at the surface is good, the 'Solid geology' maps are quite accurate, but accuracy rapidly decreases with increasing drift cover in the areas away from mineral exploitation. A drift covered rural area where the solid rock geology is complex may result in a very inaccurate 'best interpretation' Solid edition. However, much information can be gathered from river sections which cut through the soils into bedrock. A dendritic entrenched river system may provide an excellent basis for a solid map even where few natural outcrops occur on hillsides. Otherwise, where budgeting allows it, critical areas may be proved using boreholes commissioned by the geological survey.

Given that the best interpretation has been achieved on the solid geological map, what other pitfalls await the unsuspecting engineer? Perhaps the most important misconception is the use of *colour*. Geologists drawing boundaries for sedimentary formations, often use *geological time* as the basis for classification. This may involve the use of zone fossils, which are short-lived species found over wide areas and used to separate one geological *Stage* or *Epoch* from another. For instance the Namurian Epoch of the Carboniferous Period is separated from the

Westphalian Epoch by the occurrence of the fossil mollusc *Gastrioceras subcrenatum*. Rocks immediately below this level are often coloured yellow and called Millstone Grit, whilst those above may be coloured grey and referred to as Coal Measures. On the ground, however, one rock group may seem similar to the other and the difference in any engineering sense may be difficult to appreciate. A good example of this occurs on the 1:50 000 Edinburgh Solid sheet (32E). The Joppa shore exposures of NE Edinburgh, in the uppermost Millstone Grit sequence of sandstones, shales and coals, are lithologically amost identical to the lowest part of the Coal Measure sequence. Engineers may also be understandably confused by sedimentary rocks within any one colour band on the map, as it may represent more than one lithology. This is well seen on the Glasgow sheet (30 Solid), where Calciferous Sandstone Series (coloured grey) represents a group (or formation) of rocks which includes not only sandstone members, but shales, limestones and conglomerates. Similarly, the overlying Lower Limestone Group is by no means solid limestone throughout (see also Plate 1).

Some sedimentary formations do consist of one lithology with fairly consistent properties (for example the Stiperstones Quartzite in the Welsh Borderland). To sort out the inconsistences, both geologists and engineers are advised to use *Sheet Memoirs*, *Economic Memoirs* and *Regional Geological Guides* which may give the necessary details (see Appendix D).

Igneous and metamorphic rocks are more logically coloured from the engineering viewpoint, but unfortunately the detailed classification of the igneous rocks can go well beyond the requirements of the engineer. When reading the map, only the simplest of names need be quoted. On Scottish maps the key may read 'microporphyritic olivine basalt of the Dalmeny type' but the important word here for the engineer is *basalt*. He really does not need to go any further into the petrology, except to try and establish the state of weathering, jointing and strength from a subsequent site investigation. However, a geological dictionary or reference book beside the map may help to clarify the description.

1.5 Large-scale maps

Few countries publish geological maps on a scale greater than 1:10 000, the current mapping scale of the British Geological Survey (BGS). This metric scale will in future replace the marginally smaller scale 1:10 560 (6 inches to 1 mile) maps which have been in use since they became available from the Ordnance Survey in the middle of the nineteenth century. The process is likely to take many years and it is probable that engineers making reference to several different sites will experience both the 1:10 560 County sheets, some of which are lacking in contours, and the post-war 1:10 560 National Grid 10-km square sheets with contours in brown or grey. Whilst the more recent sheets are generally easier to read, and more reliable, the quality of some of the County sheets varies considerably. Stocks of most of the printed 1:10 560 maps are running out and future purchases will be photocopy prints from a master sheet.

The large scale makes them ideal for relating geological boundaries to site, since the base maps show field boundaries, individual buildings, minor water courses and on the more recent sheets, contours and even rock exposures.

Although some of the very early sheets, now only available for reference, are lacking in drift data, most maps take the form of Solid and Drift editions with indications of areas of rock, including structural data where measurements were possible.

Many of the reference maps in BGS offices have been hand-coloured and some London sheets were even published in colour. However, most published versions have the geology printed in black or brown on a grey or brown Ordnance Survey base map.

The extra space available on the maps enables additional geological information to be printed. The position of shafts and adits may be accompanied by notes giving depths to certain horizons (in fathoms or feet on older maps).

Figure 1.1 (*a*) Extract from the 1:10 560 geological map which forms part of 1:50 000 sheet 15W (New Cumnock) mapped by the Geological Survey of Scotland in the 1950s (see Plate 2). For explanatory symbols, see Figure 1.1*b*. Red lines indicate sites mentioned in Appendix A, Questions 1–3. (Reproduced by permission of the Director, British Geological Survey; Crown/NERC copyright reserved.)

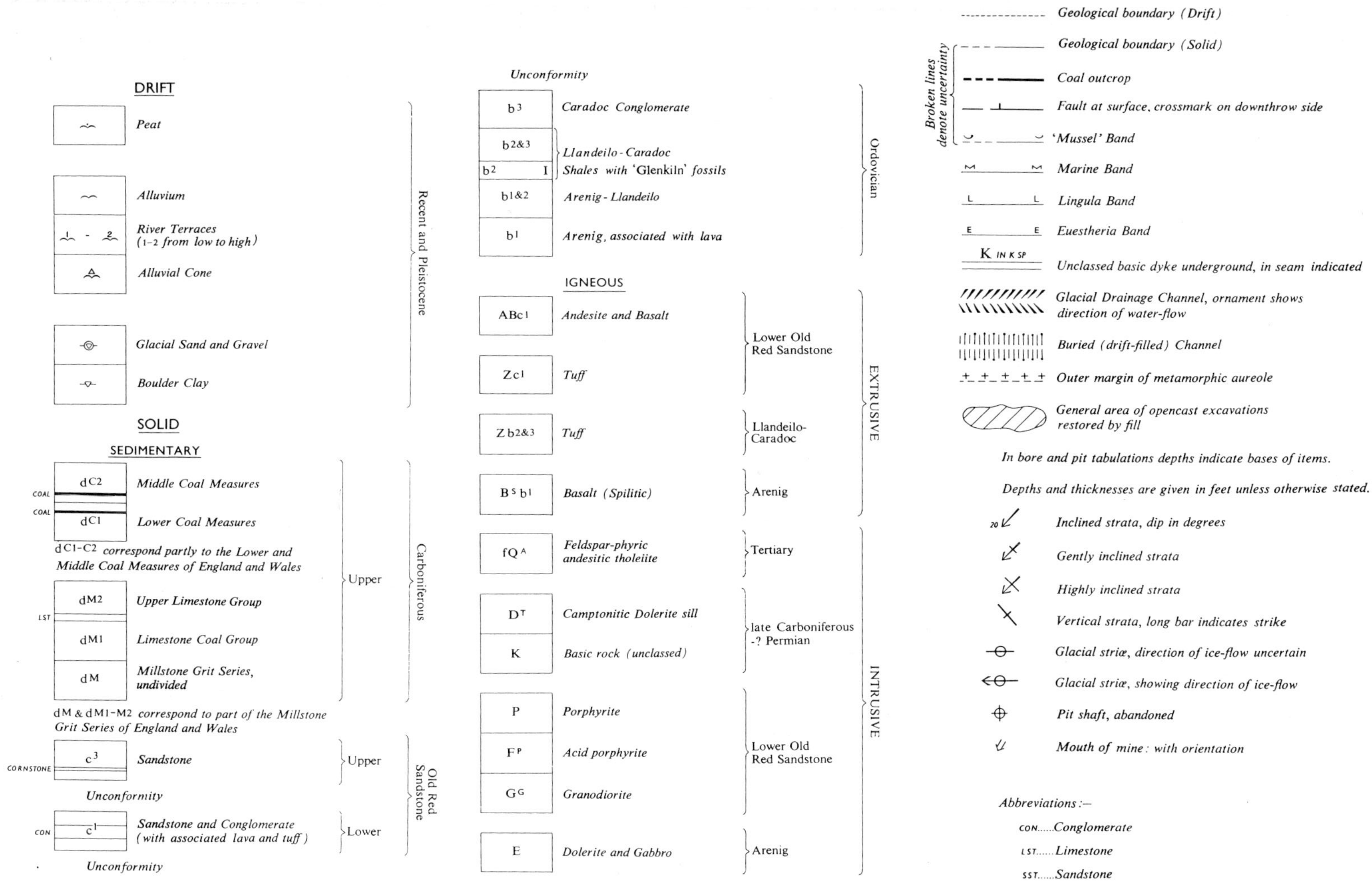

Figure 1.1 (*b*) Explanation of sedimentary rock, drift and other symbols used in Figure 1.1*a*.

Their presence obviously indicates local mining activity and may also have been used in plotting the supposed position of coal seams or mineral veins where they would intercept the surface at outcrop. Dips taken in underground workings and tunnels also help in this process and may be plotted on the map. Similarly, lithological data may be shown alongside some of the boreholes, and sequences in cliffs or river sections are also usefully listed where space allows.

Compared with smaller scale maps, more specific details of soil type are also an advantage when shown. Some geomorphological features such as the shape and width of glacial overflow channels and drumlins may appear and known depth of buried channels may be indicated.

At first sight the 1:10 000 maps can be quite confusing and it is a useful tip to have the coloured Drift and Solid editions of the smaller scale maps to hand when attempting an interpretation of the geology.

Large-scale maps form an excellent starting point for both the planning and tender stages of a civil engineering project. If the combined and detailed Drift and Solid data are used by the engineer to plan a site investigation, it is likely to be more efficient since problems may be foreseen and allowed for in the specification. Site investigation reports should also refer to the geological map coverage and any variations from expected ground conditions.

The contractor is sometimes expected to rely on borehole data and very little other information is offered. In the short time allowed for tendering, an engineering geologist should be employed to inspect the larger scale maps and present a brief report. It may be possible, for example, to identify the degree of exposure in the area of the site and gather further data. Dip symbols help not only an understanding of geological structure, but are also clear evidence that bedrock is exposed at the surface. These may be worth seeking out on site or near site in order to assess rock mass characteristics in conjunction with borehole data wherever rock excavation forms an important part of the works.

Where extensive soil cover is present and few exposures are depicted, the mapping may have been supplemented by evidence from landforms, vegetation, spring lines and perhaps auger holes. It is in areas such as this that both engineer and contractor must rely most heavily on boreholes to establish rockhead levels and both soil and rock geometry.

1.6 Background to Geological Survey coverage in the United Kingdom

The present distribution of published geological maps covering Britain is complicated due largely to the long and pioneering history of the Geological Survey of Great Britain.

It was formed officially in 1835 as a consequence of demand for a permanent record of British geology from the Geological Society of London and government agencies such as the Board of Ordnance. The Ordnance Survey had commenced the first triangulation of Britain in the 18th century and the resulting 1 inch to 1 mile County Ordnance maps first appeared in Kent, Essex and the West Country early in the 19th century and spread northeastwards. The availability of these accurate base maps enabled a small group of geologists led by Henry de la Beche to follow the Ordnance Survey by producing maps on the same scale and format, coloured geologically. The resulting *Old Series* maps were amongst the first in the world to be published and formed a systematic set numbered from south to north across England and Wales (Figure 1.2). Separate surveys were started in Scotland and Ireland about the same time.

The rapid progress northwards of this primary survey was interrupted by two important developments, beginning with the production of the larger scale 1:10 560 (6 in) Ordnance Survey maps in the mid-19th century. These provided a much better base map for field work but necessarily took much longer to complete due to the amount of detail which could be included, prior to compilation on the 1-in scale. The Old Series had reached northern England and had been initiated in central Scotland before 1860, when the larger scale mapping was introduced.

The second major development took place in the 1890s when the decision was taken to

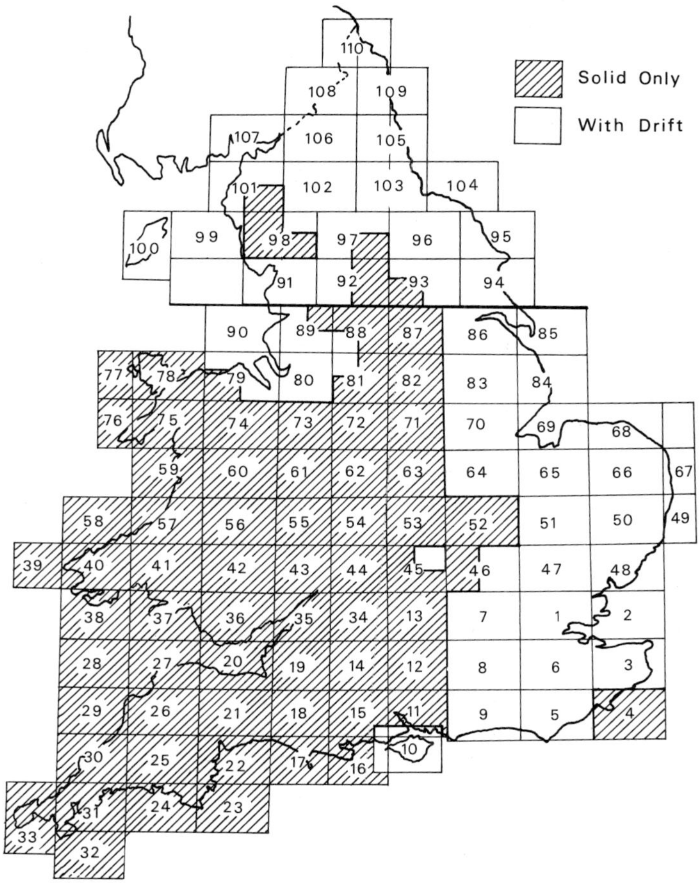

Figure 1.2 Index to 'Old Series' 1-in geological maps published by the Geological Survey of England and Wales by 1900. New sheet boundaries were adopted for Northern England following Ordnance Survey developments. After an erratic start, Old Series sheet numbering became more ordered and sheets 91–110 eventually became New Series sheets 1–73, numbered from north to south. Note: Apart from Southern England, most sheets were quartered; this is most important when ordering reference photocopies where no newer maps exist. (e.g. Sheet 98.)

abandon the labour-intensive hand-coloured Old Series for a colour-printed *New Series* following the new series Ordnance Survey maps numbered from north to south in England and Wales. New sheet boundaries were adopted to coincide with those of the new contoured edition of the Ordnance Survey county sheets (360 in number for England and Wales). These maps depict both solid geology and drift deposits, a separate Drift edition being made available wherever the abundance of superficial materials obliterated much of the solid rock. In Scotland, which retained its larger sheets, numbered independently from south to north, every sheet was intended to be produced in separate Solid and Drift editions, although some of the earlier hand-coloured sheets had been Solid only (see Plate 2).

About this time most of the work was concentrated on coalfields in Northern England and Scotland and as a consequence the most recent Old Series maps of Cumbria, Mid-Wales and East Anglia, and most of the early Scottish sheets were largely left unrevised. Two world wars not only had the effect of concentrating the mapping on economic deposits but also led to the destruction of most of the original plates by enemy bombing.

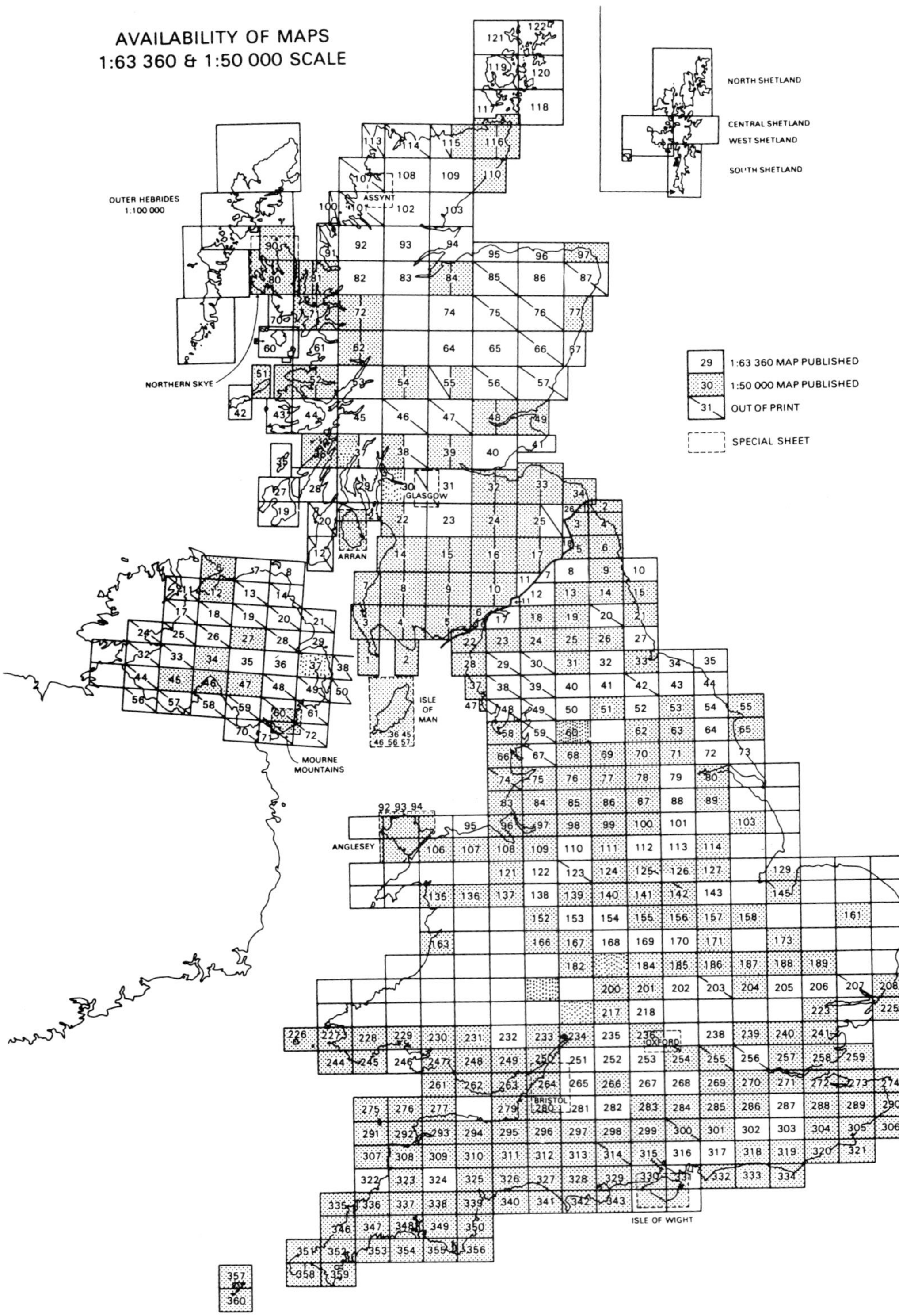

Figure 1.3 BGS map coverage, 1990.

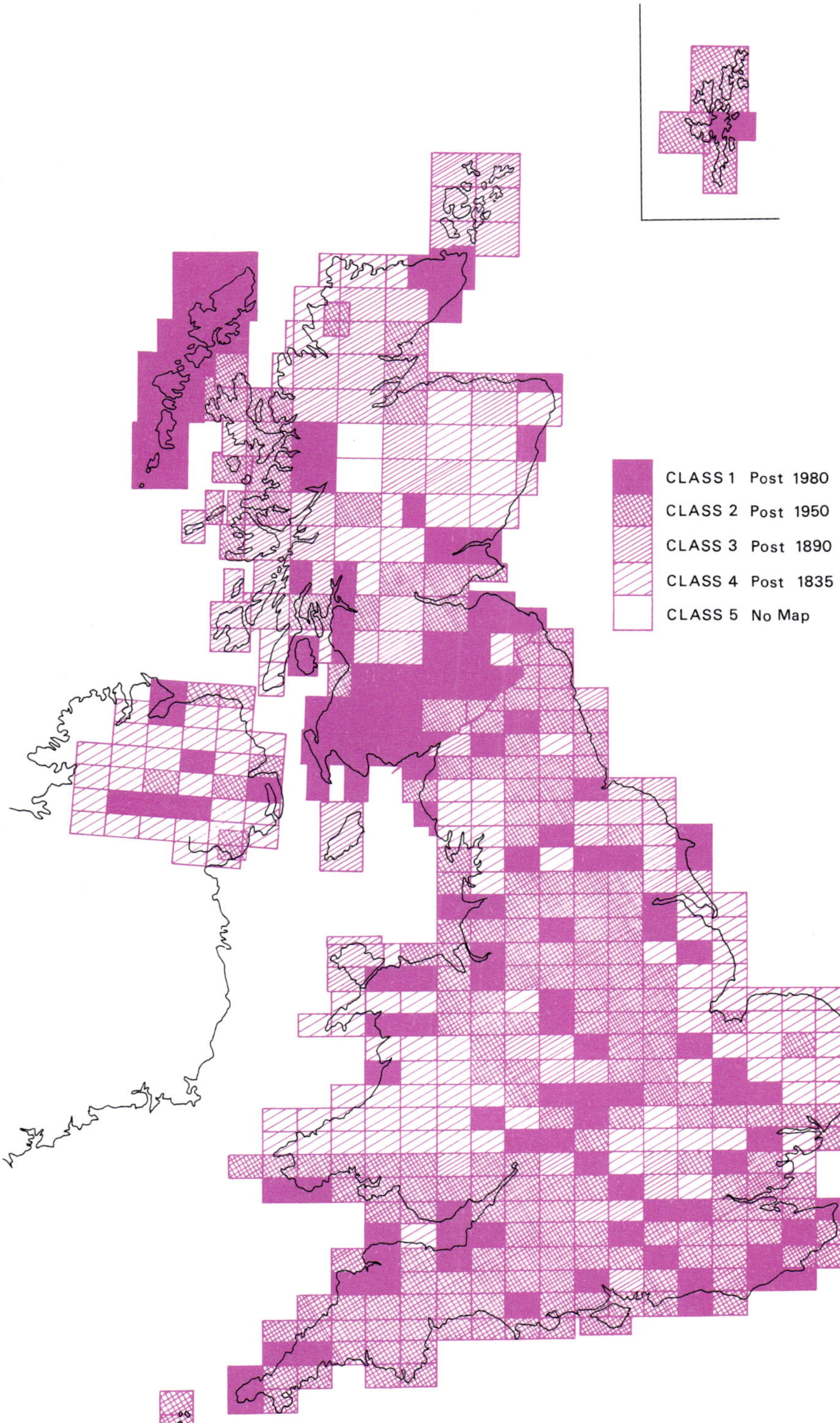

Figure 1.4 Geological Survey sheets of Great Britain categorised into classes 1–5 according to date of mapping.

The post-war effort continued the replacement of many older sheets and included the introduction of maps on a scale of 1:50 000. These are gradually replacing the older 1:63 360 Old and New Series maps (Figure 1.3). All field mapping is now concentrated on a scale of 1:10 000 but the National Grid Sheets at this scale will not be published in future since dyeline copies from transparent masters will be available to order.

As a result of the long history of mapping there is a variation in the quality of geological map coverage and reliability throughout Britain. An engineer should be prepared to deal with the following categories of map presently available but which are steadily being improved. They are listed here in order of decreasing general reliability, although some of the older maps are of excellent quality.

Class 1. Recent 1:10 000 (or 1:10 560) National Grid re-surveys accompanied by metric 1:50 000 Solid and Drift sheets. These may incorporate borehole and geophysical evidence from site investigations and are normally of excellent quality and usefulness (e.g. England and Wales sheet 80 Kingston upon Hull). Post 1980

Class 2. National Grid 1:10 560 primary or re-surveys accompanied by 1:63 360 Solid and/or Drift sheets, sometimes enlarged to 1:50 000 without revision (e.g. Scotland sheet 32E Edinburgh). Post 1950

Class 3. Published older 1:10 560 County 6-in sheets dating back to 1860 accompanied by older New Series sheets at 1:63 360. Colour-printed since the turn of the century (e.g. current coverage of England and Wales sheet 268 Reading Drift is dated 1904). Mainly Post 1890

Class 4. Out of print Old Series 1-in and early New Series hand-coloured library reference maps. Solid editions only at 1:63 360. Not updated (e.g. E & W Sheet No 29 Keswick, Old Series 101 SE). Post 1834

Class 5. No geological map. Primary surveying being pursued by BGS (only a few sheets in the Central Highlands left in this category).

It is clear that sites within Class 1 coverage will have the advantage of the most up to date metric base maps and will include the latest mapping techniques and data from the results of previous site investigations. However, as the class numbers increase the data become progressively less satisfactory, with mapping scales being reduced and only solid geology shown (Figure 1.4).

Figure 1.3 shows current BGS medium scale coverage including special sheets and some 1:100 000 maps of the Outer Isles. For areas such as central Cumbria, and East Anglia it may be necessary to refer to the newer but much smaller scale *Universal Transverse Mercator* (UTM) series of maps which cover both the mainland and continental shelf areas and are produced on sheets dealing separately with solid geology, quaternary geology and sea bed sediments (Figure 1.5). Corresponding gravity and aeromagnetic maps of the same areas are also available as detailed in the BGS *Catalogue of Printed Maps.*

Certain areas of classical British geology are covered by 1:25 000 coloured maps, which have most of the advantages of the 1:10 000 sheets (including field boundaries), but are much easier to read. Unfortunately few sites will occur in such areas. Other more specialised maps of value to engineers will be dealt with in Chapter 4.

It should be appreciated by British civil engineers that the above situation is something of a luxury since no other country appears to use a scale of 1:10 000 for its primary field surveys and many large countries such as Australia and Canada only publish maps on scales of 1:250 000 or even smaller. Few countries produce geological maps with the amount of structural detail which appears on BGS maps and surprisingly few have scaled legends for showing the thickness of formations.

An exception is the United States Geological Survey which despite the size of the task produces detailed geological and topographic maps at 1:24 000 and 1:62 500 (the $7\frac{1}{2}$ min and 15 min quadrangle maps), calling for 30 000 sheets. Over 1500 have been completed to date. Even the

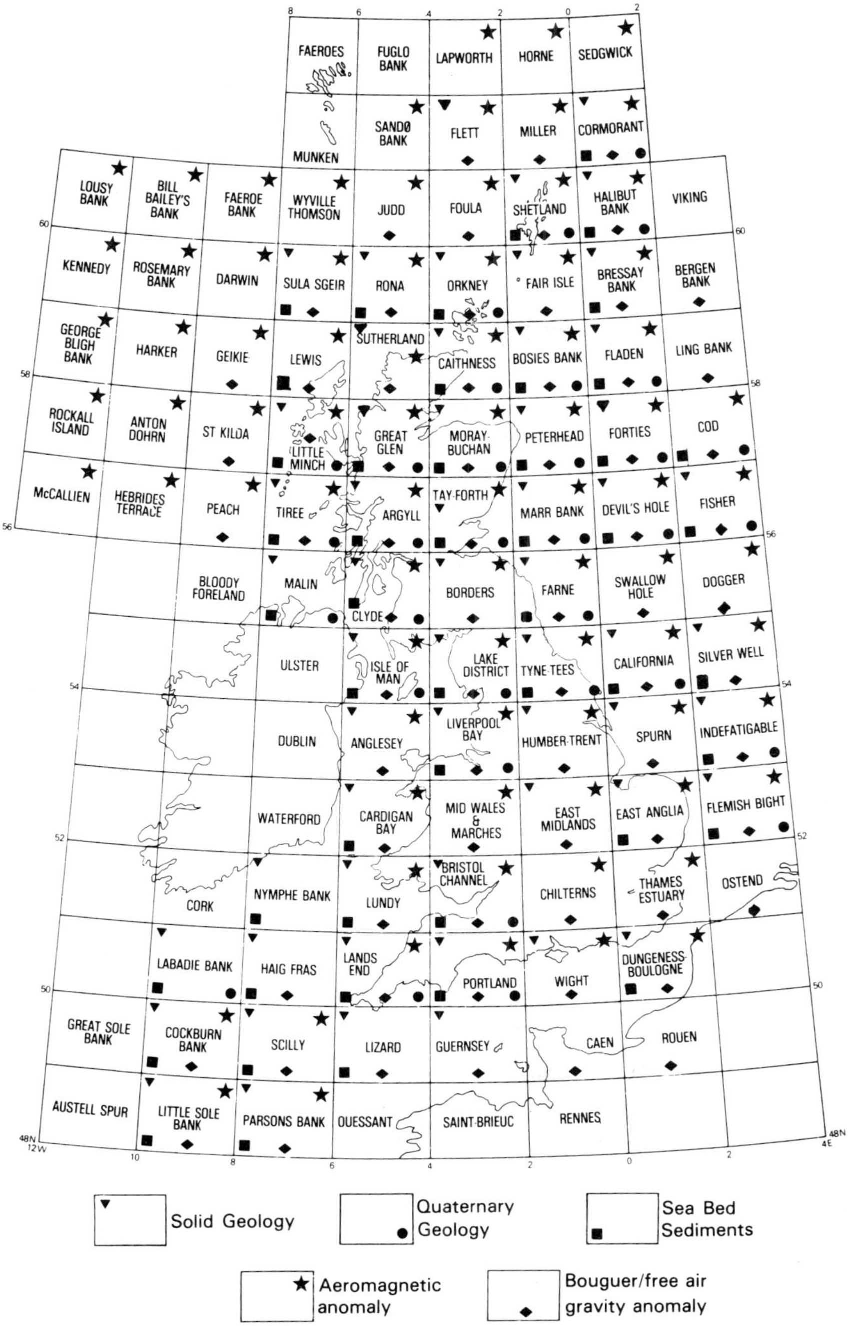

Figure 1.5 1:250 000 scale coverage of the United Kingdom: BGS availability of maps, 1990 (UTM series).

older maps show a very high standard of cartography with details of both solid and drift clearly shown. Some of the more recent maps give extra engineering-orientated data relevant to the particular areas covered (see Plate 4*a*). The original folios of the *Geologic Atlas of the United States* started in 1894, were continued in the GQ Series of maps begun in 1949.

Chapter 2 Reading geological maps

2.1 Outcrop patterns

It is natural to start with the most simple patterns produced by outcropping rock and gradually increase complexity. Initially, then, only sedimentary strata with constant thickness will be considered.

2.1.1 *Horizontal strata* (Symbol +)

The simplest geological maps are those where there is little surface relief and where the strata are horizontal. The result is a single formation (or colour) over the whole map. This situation occurs only rarely, but in Britain is approached in NE Caithness and East Anglia.

Apparent complexity may be demonstrated wherever valleys and hills cut through the horizontal strata. The elaborate shapes depicted on geological maps of the North Pennines of England and Grand Canyon, Arizona are out of all proportion to the actual simplicity of the geology—such is the overwhelming influence of topography!

Where strata are horizontal, the geological boundaries run parallel to topographic contours so that interpretation is actually very easy, provided topographic contours are shown. Thicknesses may also be easily calculated from the heights above sea level of the top and base of the unit to be measured (Figure 2.1*a*).

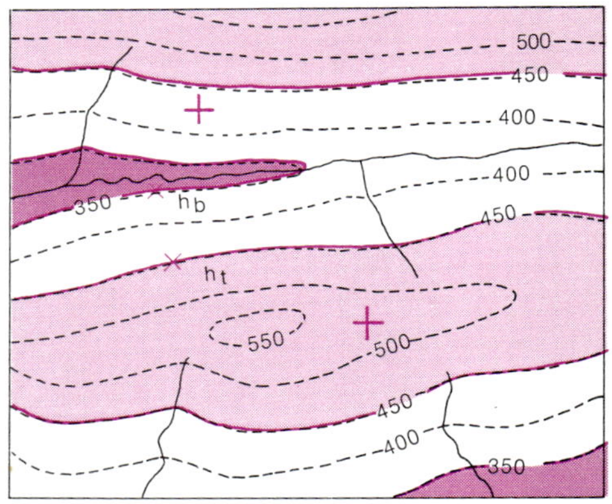

a.

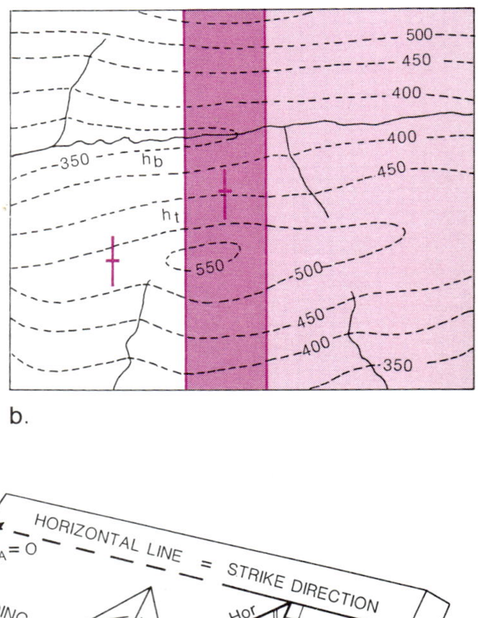

b.

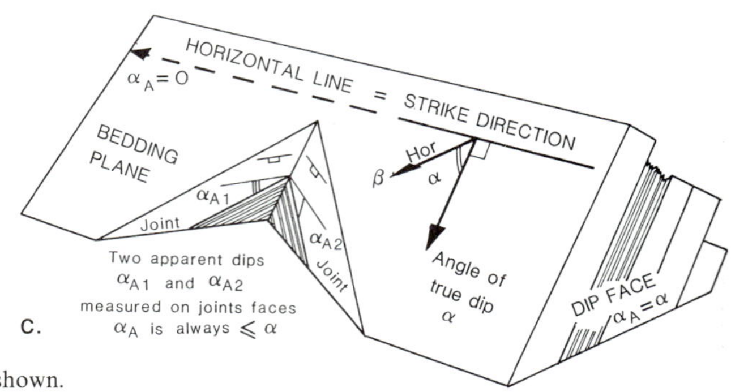

c.

Figure 2.1 (*a*) Map of horizontal strata with symbol shown. $T_s = h_t - h_b$. (*b*) Map of N–S striking vertical strata with symbol shown. T_s is given by width of outcrop. (*c*) Block diagram to show true bedding dip α, apparent dip α_A, and direction of dip β.

2.1.2 *Vertical strata* (Symbol —+—)

Where strong tectonic influence upturns sedimentary layers, they may intersect the surface vertically. Similarly, some igneous intrusions such as dykes may have vertical boundaries. If this occurs on a geological map, the boundaries may be straight lines and certainly appear to be independent of topography. The boundaries cut across contours and are not influenced by them in any way. The thickness of units which have a vertical disposition is equal to their outcrop widths and can be measured directly using the appropriate map scale (Figure 2.1*b*).

In general engineering terms, the presence of steep or vertical strata may indicate rapid changes of lithology across the site, especially when compared with a level site on horizontal strata.

2.1.3 *Inclined strata* (Symbol $\rightarrow\alpha$, preferred symbol $\underset{\alpha}{\top}$)

The dip of a rock plane is defined as the maximum angle of inclination of that rock surface measured from the horizontal. For bedding planes it is therefore measured using a clinometer-compass perpendicular to the strike or horizontal line on the bedding surface (Figure 2.1*c*) and its orientation is recorded as being the angle from the horizontal (α) and the bearing (or direction clockwise from north) of that angle (β).

Since the outcrops associated with inclined strata are deflected across hills and valleys, they produce informative and sometimes complex patterns on geological maps, depending on dip angle and topography.

Inclined strata outcropping in flat country. In a flat area, a succession of different lithologies is represented by parallel-sided bands on the map as each bed in turn comes to the surface. The younger strata outcrop successively in the direction of dip and any increase in width of outcrop indicates lowering of dip angle (Figure 2.2*a*) (e.g. SW corner of BGS sheet 108(S) for the Triassic rocks of the Cheshire Plain).

Inclined strata outcropping in hilly country. The amount of outcrop deflection in hilly country depends on several factors:

(1) the angle of dip of the strata (α);
(2) the direction of dip of the strata (β);
(3) the steepness and form of the topography.

Fortunately there are some relatively simple geometrical rules which can be followed for general interpretation, the most important of which is the *V-shape rule* which states that:

In a valley, V-shaped outcrops point approximately in the direction of dip.

This enables the map reader to determine which way the beds are dipping by looking at the relationship between the geological boundaries and the topographic contours. A further general rule is that *lower angles of dip* produce more *acute V-shapes* across hills and valleys (Figure 2.2*b*, *e*) whereas *high angles of dip* result in only slight *obtuse V-shapes* (Figure 2.2*d*).

However, these are only general rules and it is necessary to be aware of certain variations and exceptions to be able to read any geological map.

(1) The width and depth of valleys will also control the acuteness of V-shaped outcrops so that they may become broad U-shapes in some wider valleys (Figure 2.2*c*).
(2) The gradient of the valley floor is also a factor which can produce more obtuse or acute V-shapes and may also lead to an exception to the V-shape rule. If the valley floor is steeper than, but in the same approximate direction as, the bedding dip, it will form a V-shaped outcrop pointing in the opposite direction to the dip. This is one of the reasons why there are isolated outcrops of older strata (*inliers* Figure 2.2*f*) and younger strata (*outliers* Figure 2.2*g*) in many hill areas where gently inclined rocks occur.
(3) Width of outcrop can change suddenly despite the fact that stratum thickness may remain constant. The lower the angle of dip relative to slope the greater the outcrop width (W), except where topographic slope is

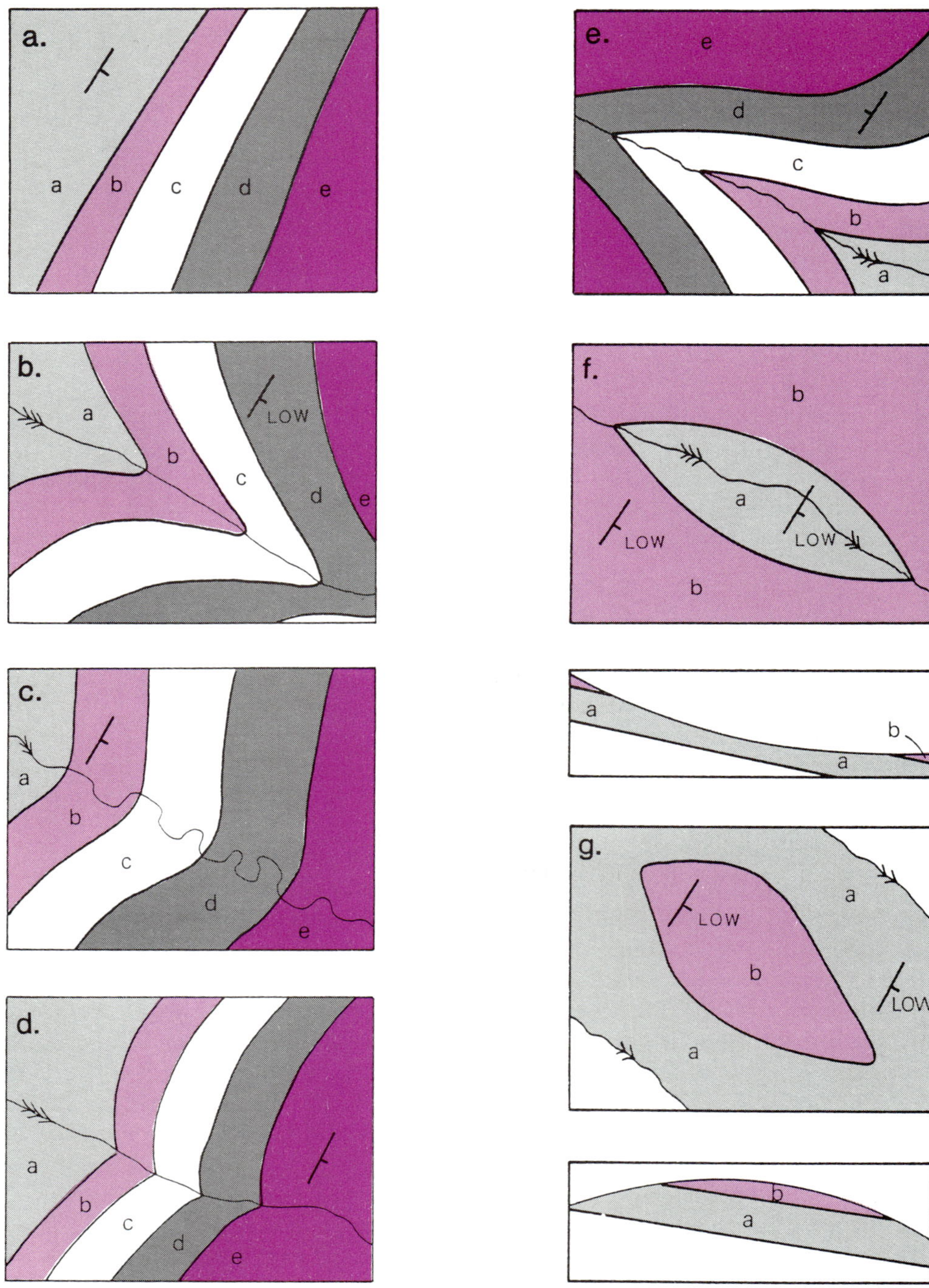

Figure 2.2 Sketch maps illustrating V-shape rule for dipping strata, and its exceptions (e–g) for low dipping strata: (*a*) Flat topography; (*b*) Low dip, narrow valley; (*c*) Low dip, wide valley; (*d*) Steep dip, narrow valley; (*e*) Low dip up valley*; (*f*) Map and section, inlier; (*g*) Map and section, outlier. Note: Strata a (pale grey) are oldest and strata e (dark red) are youngest. (*without contours or a dip symbol, this could represent horizontal strata, or a shallow dip down valley where gradient is steep.)

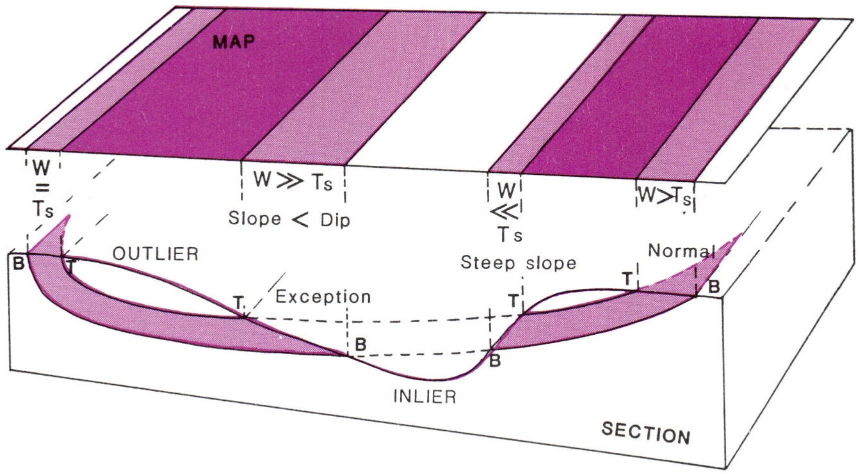

Figure 2.3 Block diagram illustrating the dependence of width of map outcrop on dip and slope and explaining top (T)-base (B) configurations.

very steep. On steep slopes the width will decrease significantly (Figure 2.3).

(4) Recognition of base and top of lithological boundaries on the map may be helped using the following rule: when looking down dip, the base lies nearer than the top, *except* when the slope is steeper than the angle of dip *and* in the same direction (Figure 2.3).

Appendix A contains problem maps on which these qualitative rules may be practised prior to any quantitative analysis.

2.2 Quantifying planar structures

To extend the above qualitative approaches to reading simple geological structures, more quantitative geometrical methods are available. Contours drawn for example to represent the ground surface on a map, may also be constructed for a bedding plane below ground. The resulting *stratum* or *structure contours* are equivalent to strike lines traced out at specific heights above or below sea level, usually at the same interval as topographic contours.

The dip of the strata controls the spacing and shape of the structure contours drawn on it. On a map these contours often form simpler shapes than topographic contours. They are drawn for a specific surface whose outcrop at ground level is defined by the intersection of the two sets of contours, because where equivalent values of topographic and structure contours cross each other, an outcrop of that particular boundary surface must occur. The three interdependent variables are:

(1) outcrop pattern;
(2) topographic contours;
(3) structure contours.

Thus given any two of these, the third can be plotted. For example, if the outcrop of the base of a planar structure cuts a single topographic contour in several places, the intersections may be marked with a cross. Joining the locus of the crosses will form a single structure contour for that base (Figure 2.4).

For simple uniformly dipping strata, structure contours are straight, parallel and equally spaced (Figure 2.5*a*). For a changing dip angle they are straight and parallel but not equally spaced (Figure 2.5*b*), and for a changing angle *and* direction of dip they are curved and not equally spaced (Figure 2.5*c*). In nature the third case is most common, but for engineers working on small excavations in an area of simple

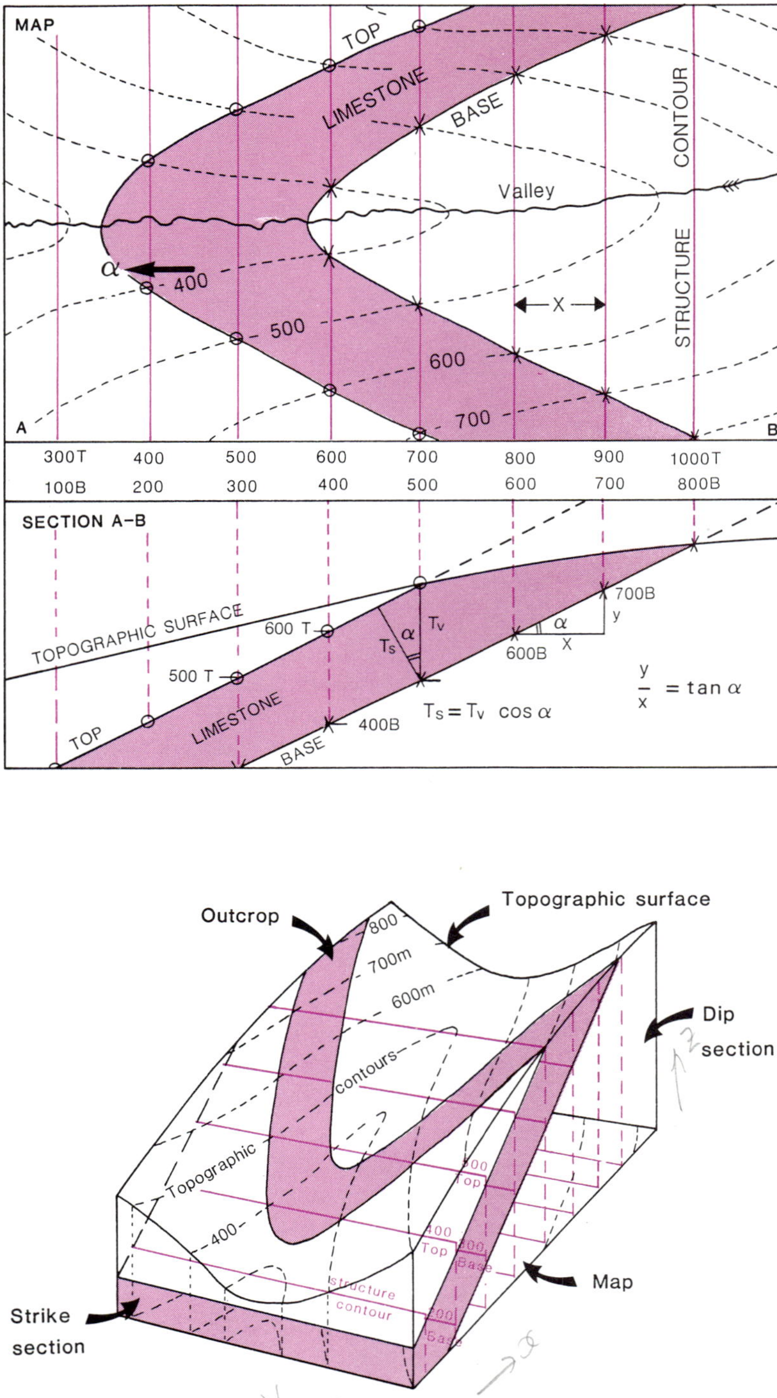

Figure 2.4 (*a*) Map and section showing the interdependence of topographic contours, stratum contours and outcrop pattern. y/x = tan α, where α = angle of dip, x = distance between structure contours and y = contour interval; $T_s = T_v \cos \alpha$, where T_v = vertical thickness and T_s = stratigraphical thickness. (*b*) Block diagram showing the interdependence of the topographic contours, structure contours and outcrop pattern as shown in (*a*).

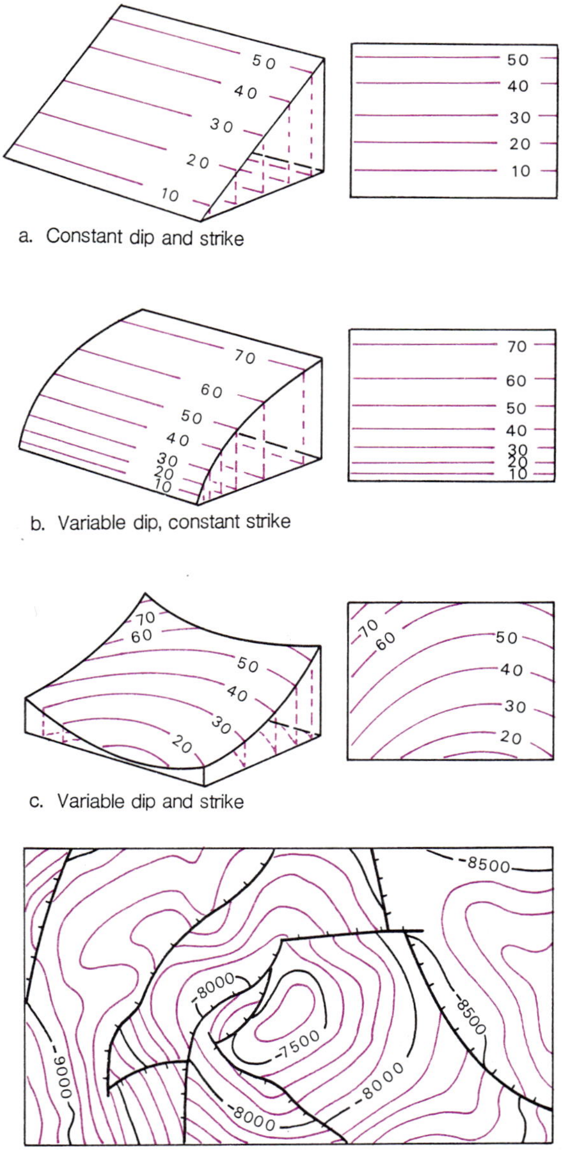

a. Constant dip and strike

b. Variable dip, constant strike

c. Variable dip and strike

d. Structure contours on a typical North Sea oilfield.

Figure 2.5 Structure contours.

geology, the structure contours may be a useful tool for predictive extrapolation or interpretation. In coal-, gas- and oilfields, sophisticated computer methods of plotting structure contours may become necessary in order to understand the three-dimensional nature of important large-scale and sometimes complex structures (see Figure 2.5*d*).

2.2.1 *Calculation of angle of dip*

Many geological maps are published without dip symbols. Structure contours constructed from the combination of outcrop boundaries and topographic contours enable average dips to be measured. It only requires two structure contours to enable the average dip between them to be calculated. If the horizontal distance between structure contours of interval y metres is measured from the map as x metres then (see Figure 2.4)

$$\text{The average dip } \alpha^\circ = \cot y/x \qquad (1)$$

Numerous opportunities to put this into practice can be found in the maps in Appendix A.

2.2.2 *Calculation of thickness of inclined strata*

There are several methods of calculating the thickness of inclined strata. In civil engineering the vertical thickness (T_v) may be just as important as the stratigraphical or true thickness (T_s) but either may be calculated from the following relationship (Figure 2.6*a*)

$$T_s = T_v \cos \alpha \qquad (2)$$

The difference becomes significant for high values of α but if $\alpha < 10^\circ$, T_s is very similar to T_v.

Dip methods (measurements in the direction of dip).

For flat country: the thickness of the inclined strata may be obtained by measuring the outcrop width (W) in the direction of dip. Then (Figure 2.6*a*)

$$T_s = W \sin \alpha \qquad (3)$$

and the borehole thickness

$$T_v = W \tan \alpha \qquad (4)$$

For hilly country: the topography will have to be allowed for unless structure contours for the *base* and the *top* of the unit to be measured can be compared at one place on the map (Figure 2.6*b*). This is the simplest method but if structure contours are not readily available, the following method is used. If W is the width of the outcrop in the direction of dip, h_t is the height above sea level of the top of the unit and h_b is the height above sea level of the base of the unit, so that $h_t - h_b = H$ (the height difference). Then (Figure 2.6*c*)

$$T_v = W \tan \alpha \pm H \qquad (5)$$

Use $+H$ where the dip is opposite to the slope (remember by using *opposites attract*) and use $-H$ where the dip and slope are in the same direction, i.e. where h_b is down dip.

Other graphical methods involving the drawing of sections are also valid but take longer to calculate. T_v is measurable from any apparent dip section but T_s can be measured directly only from a true-scale dip section.

The strike method. A simple method of finding T_v for relatively thin units is known as the strike method since it utilises the fact that all sections drawn parallel to the *strike direction* contain apparently horizontal strata. Thus a *single* structure contour cutting *both* base and top

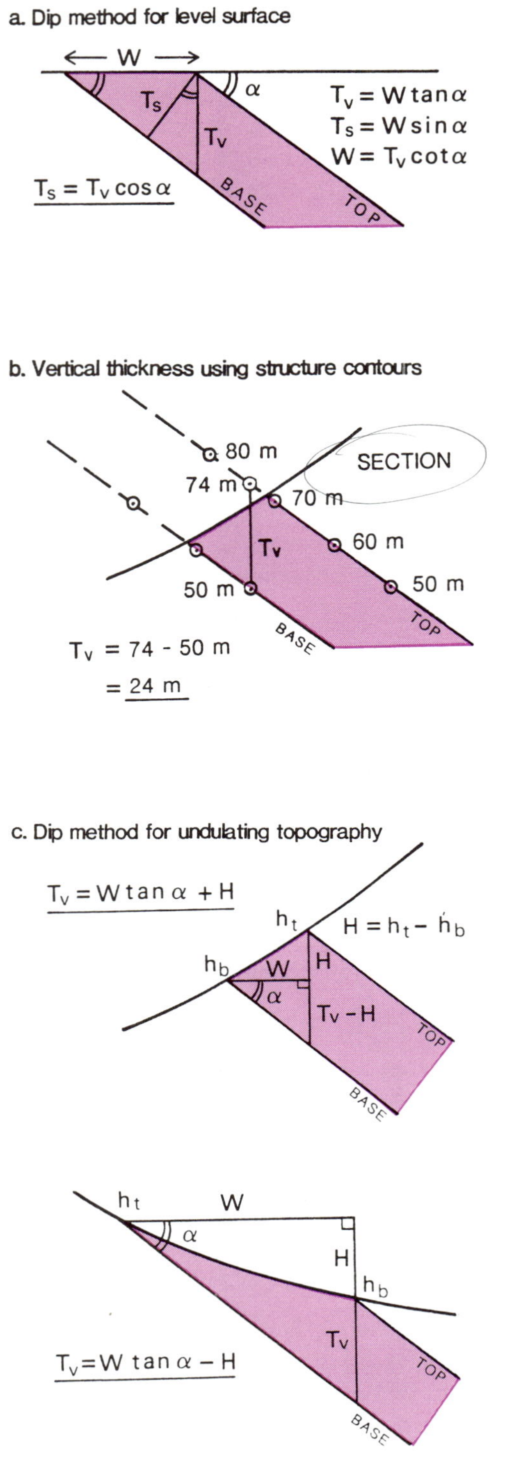

Figure 2.6 Dip methods for calculating thickness of strata.

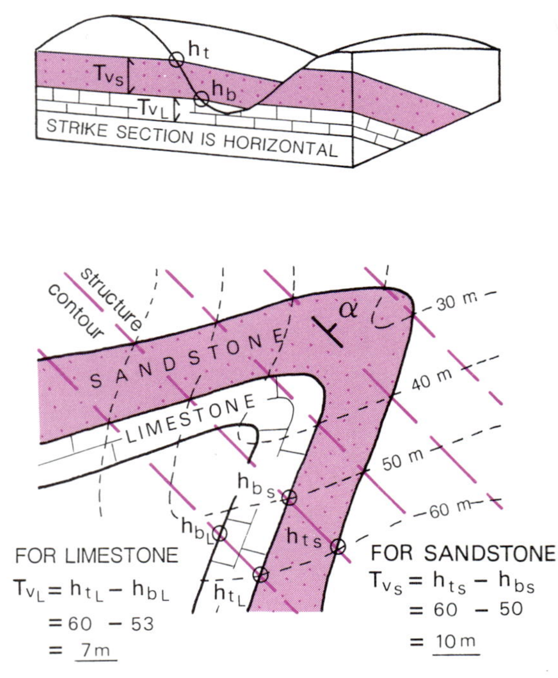

Figure 2.7 The strike method for estimating vertical thickness, T_v, using structure contours.

boundaries is all that is needed for the calculation. h_t and h_b are measured or estimated at the intersections of the chosen structure contour and the difference between them is T_v for the rock unit in question. Thus the vertical thickness using the strike method is obtained from

$$T_v = h_t - h_b \tag{6}$$

Figure 2.7 illustrates its simple application for two units, one measured and one estimated (see also Problem Maps A.1–A.7 in Appendix A).

Unfortunately, the strike method is not always applicable, especially when the unit to be measured is of considerable thickness, so that a single structure contour does not intersect both top and base of the unit (try map A.7).

2.2.3 *Three-point solutions*

Where simple geological structures are present it may be possible to predict the position of a certain rock layer such as a coal seam by finding its reduced level at a few points.

The *minimum* number of data points to fix the orientation of a plane is three; hence the name three-point solution. In practice the more numerous the data points, the more accurate the prediction. However, to illustrate the principle the reader is referred to Figure 2.8, where three boreholes have been put down to find the underground position of the top of a coal.

Once the *reduced level* of the top of the coal has been worked out for each borehole, a triangle joining the boreholes can be proportioned to give crossmarks, in this instance, at every 5-m

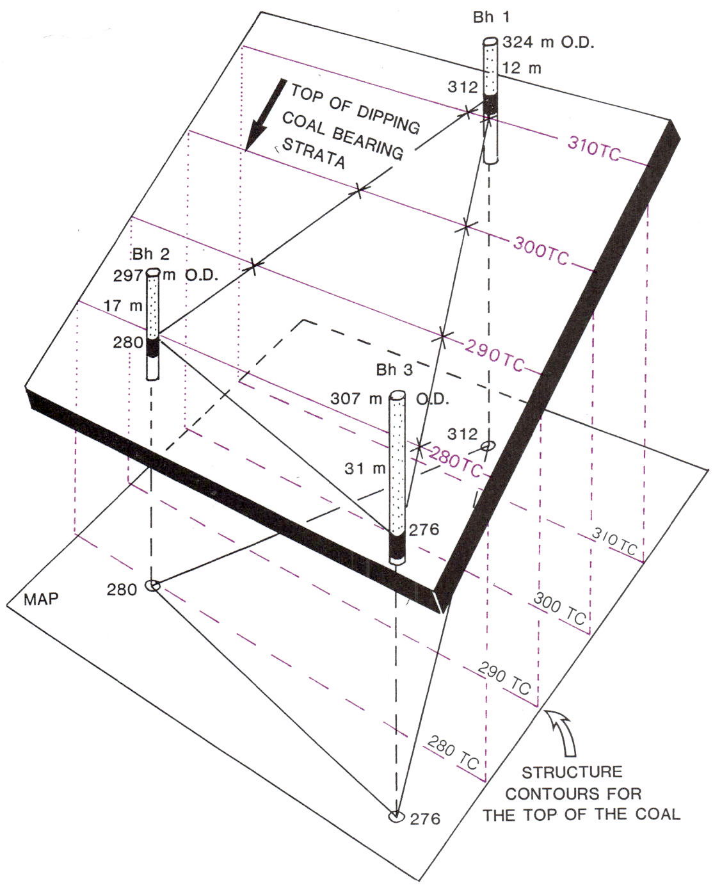

Figure 2.8 Block diagram to illustrate the construction of a three-point solution.

structure contour position. For a dipping planar coal seam the contours may then be drawn on the map by joining points of equal height on the triangle. The resulting structure contours may then be used in conjunction with topographic contours to predict outcrops, and work out the angle and direction of dip.

Of course it is not essential to use only boreholes. Any combination of hillside or quarry exposures and/or pits and boreholes with reduced levels may also be used. At an excavation site it may involve three or more surveyed-in boundary levels. However, assumed planar conditions are unlikely to extend over large areas and further boreholes should be used to build up a network of triangles or quadrangles which may help to establish a more accurate three-dimensional shape, taking into account recorded faults, suspected faults and folded strata (Figure 2.9*a*) (try problem maps A.2*a* and *b*).

Sophisticated structure contour plans may be computed from borehole and seismic results to predict the structure of opencast sites and oilfields (Figure 2.5*d*) or simply to estimate the position of particular rock units in unexposed ground when producing a Solid geological map.

Engineers working on a small excavated exposure at a known elevation may, however, find that the *direct measurement* of dip angle and direction using a *clinometer* and *compass* is the most effective method of estimating structures over a small area. The angles obtained, if representative of the bedding dips over the area, may be converted into gradients from which structure contours may be constructed.

Prediction of the geology at a level not yet uncovered is an important part of engineering geology. Dips or structure contours obtained either from geological surveying, maps or boreholes can be used to predict rock conditions at the base of a proposed large excavation or at a level where benching of a hillside is to be carried out. (Two simple examples are given in Figure 2.10 and problem map A.5). Where the top and base of a stratigraphical unit forms a wedge rather than remaining constant in thickness, points of equal thickness may be plotted as *isopachs* (or isopachytes). More commonly in site

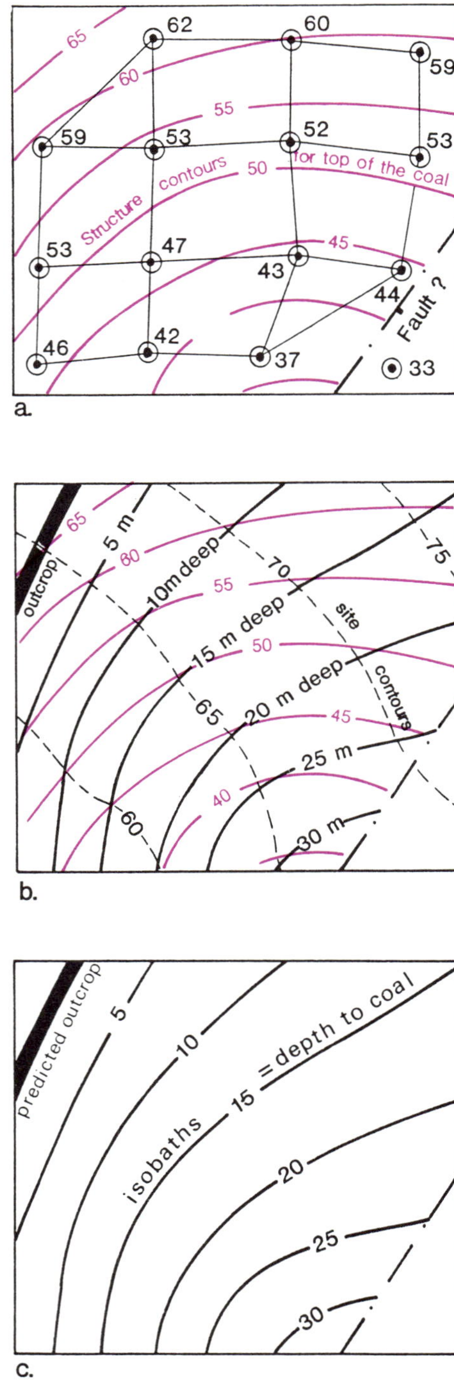

Figure 2.9 Multipoint contouring: (*a*) Reduced levels in boreholes to the top of a coal seam are used for multipoint construction of structure contours; (*b*) Combination of structure contours and site contours form depth contours (isobaths) to the top of the coal; (*c*) Isobath map.

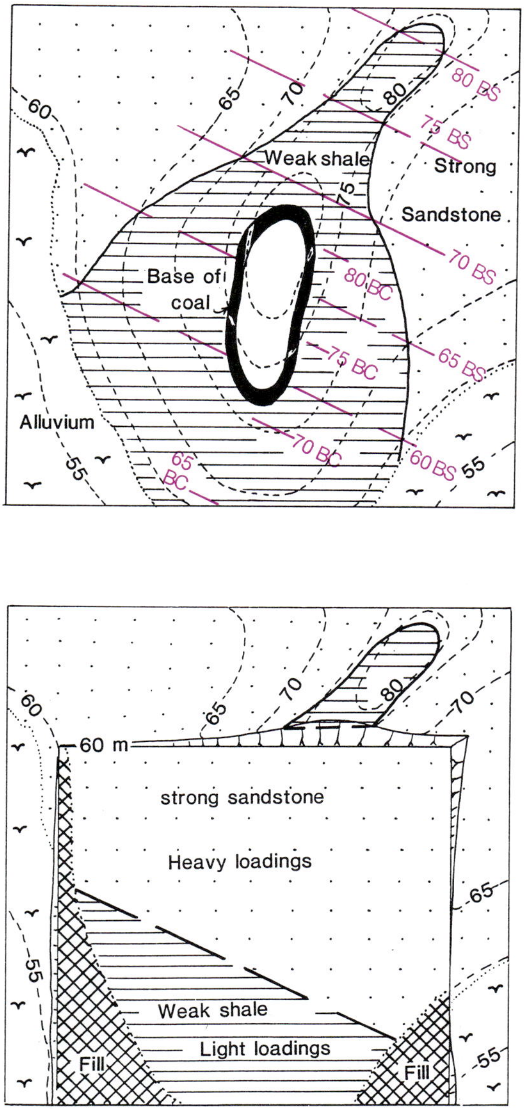

Figure 2.10 (*a*) Geology of a site to be benched (with relevant structure contours). BC = structure contours for base of the coal, BS = structure contours for base of the sandstone. (*b*) Predicted distribution of materials on a level bench site at 60 m above sea level.

investigation work where boreholes provide most of the data, points of equal *vertical* thickness, or *isochores* are plotted for a particular unit. The most valuable data for the engineer involve the depths from the ground surface to rockhead or some other critical surface. When contoured from borehole data, these are known as *isobaths* or depth contours, as distinct from structure contours which give heights above (or below) sea level for the same surface. Whilst isobaths may be constructed directly from borehole data, the intersection of topographic and structure contours provides a more realistic set of results (Figure 2.9*b*, *c*).

2.3 Structural complications

Unfortunately, unless sites are small or in areas of relatively undeformed rock, the simplicity of the patterns produced by the strata described above may grossly under-represent the actual geology.

In many areas the outcrop patterns on geological maps are complicated as a result of the response of the rocks to earth movements. Given geological time they may be folded and faulted to various degrees of severity. It is common in many orogenic belts, such as the Highlands of Scotland, for rocks to have been repeatedly folded and faulted, giving extremely complex outcrop patterns beyond the scope of this book. On the ground, further complexity may be present in the form of numerous sets of discontinuities such as minor faults, joints, cleavage and bedding planes, which are not apparent on many geological maps.

2.3.1 *Folded strata*

Given geological time, sufficient confining pressure or increase in temperature, even competent rocks may respond to prolonged stresses by buckling, flexing or shearing to form folds in the earth's crust. The subsequent denudation processes result in outcrop patterns which may be readily recognised on geological maps. A fold will normally consist of two limbs separated by a hinge zone. This leads to a repetition of strata at the surface on opposite sides of the fold axis with, in the simplest cases, younger rocks in the centre of downfolds (*synclines*) and older rocks in the core of the upfolds (*anticlines*). Symmetrical folds with near-horizontal axes form parallel outcrops with

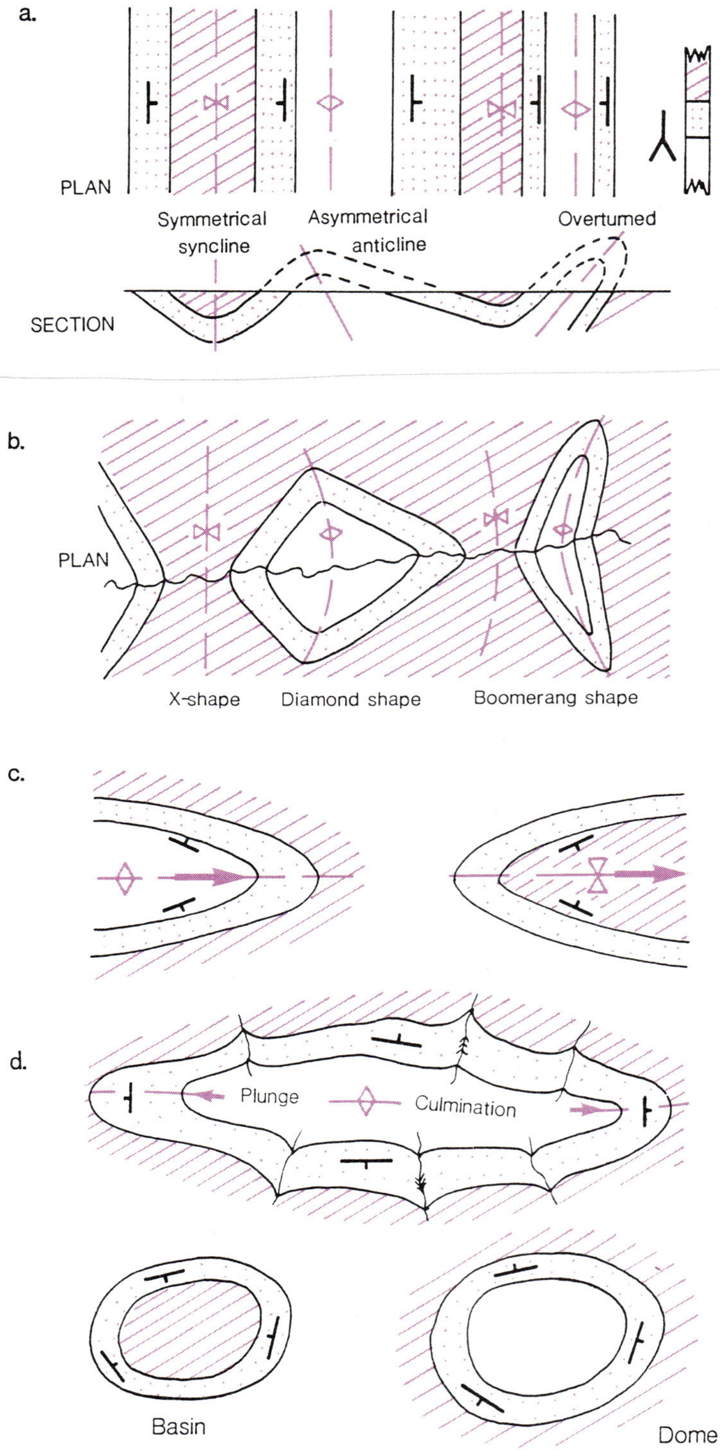

Figure 2.11 Folds on maps: (*a*) Folds—no plunge, parallel axes, level ground, note repetition; (*b*) Folds—as (*a*), but crossing a valley; (*c*) Plunging anticline and syncline; (*d*) Domes and basins—eroded elongated dome.

symmetrical repetition about the axis. Asymmetrical folds or a sudden steepening of topography may cause outcrops on one side of the axis to be much wider than those on the other (Figure 2.11*a*). V-shaped outcrops in valleys clearly show this change and if the topography involves deep valleys then the disparate V-shapes may join up over the hinge to form *diamond* or *X-shaped* outcrops on the map.

Increasing asymmetry of dips on the limbs results in D-shaped, K-shaped, or boomerang-shaped outcrops (Figure 2.11*b*) but these appear more frequently on geologist's problem maps than on the ground, since most valleys tend to follow the trend of folds and only occasionally cut across them (and then perhaps in gorges).

Fold axes commonly plunge into the ground so that the hinge zone forms a curved outcrop eroded to the topographic surface. In horizontal non-plunging folds, the structure contours may be sub-parallel on the limbs, but wherever plunge brings the limbs together over a hinge, structure contours also curve around the hinge. A very useful fact for interpreting shape is that *anticlinal* limbs *converge* in the direction of plunge (Figure 2.11*c*). In general elongated domes (or culminations) form elliptical outcrops (Figure 2.11*d*), well illustrated for example on the geological maps around Bristol (especially the Bristol Special 1:63 360 sheet) and the Wealden Dome on the 10 mile, South Britain sheet. Engineering sites are often too small to contain large scale fold patterns, but in certain tunnel projects and some reservoirs, they can be very important in controlling groundwater flow and, of course, lithological distribution and ease of excavation (try problem map A.3). Reference to published geological maps even on a small scale may be the main indication that such problems exist and give advance warning as to the type of site investigation to be specified.

An example may be cited from the rock excavation problems that arose during the construction of the Kariba Power Station in Zambia during the early 1970s. Both the engineer and the contractor appear to have been unaware of the likely presence of folded weak mica schist units within the gneiss of the underground machine hall, despite recommendations made by the Zambian Geological Survey who had mapped surface exposures.[1]

2.3.2 *Faulted strata*

Brittle reaction to deformation sometimes involves joints and small faults which cannot easily be depicted at the scales used on most published geological maps. However, when displacement is large enough to affect the outcrop pattern significantly, they are quite often shown on the map. Actual fault exposures are rare since they are normally zones of weakness, more easily eroded than the surrounding rocks, and are often covered by river alluvium. If, however, the fault zone is mineralised with quartz or some other hard material, it may stand as ridges rather than form a valley. On most maps faults are shown as thicker black, red or blue lines, often dotted or dashed to indicate their uncertain position.

A warning: Fault frequency on geological maps merely reflects the extent of rock exposure, or perhaps the presence of mining, where faults are encountered and plotted underground.

Although faults (particularly active faults) may have a major effect on civil engineering works it is a common misconception among student engineers that their presence spells disaster! As mentioned above, the fault zone may even be stronger than surrounding bedrock. Normally, however, inactive faults offer little direct threat unless they affect groundwater flow, weathering grade, excavation stability or occur in zones where high loadings are to be imposed. They may, however, become re-activated by underground mining or by water, gas or oil extraction or fluid injection. Further reference to their inclusion on geological sections will be made in the next chapter.

The origin, terminology and classification of faults is documented in many basic geological texts and so it is only their effect on geological map outcrop shapes which is considered here. When reading published geological maps, the effect of faults is most obvious on Solid editions. Trends of various fault patterns and the type of

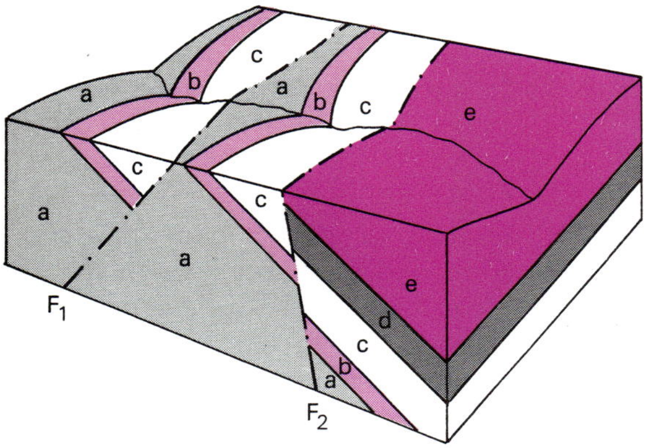

Figure 2.12 Block diagram illustrating the repetition of strata when a normal strike fault, F_1, cuts beds a, b and c dipping in the opposite direction. When the dip of the strata and the normal fault F_2 dip in a similar direction, some units (e.g. unit d) may be cut out and difficult to detect on a geological map.

displacement may be readily recognised by good map observation. In this section a qualitative approach will again be presented first, followed by methods of quantifying fault throws.

Fault data from published maps. Faults running with sub-parallel trends may be related in age and type. Congruent sets with orientations differing by 30° or 60°, but also related to the same stress field, may be present. The age may be estimated by the cross-cutting of beds or dykes of known age and by the relationship of the fault with unconformable strata, or later igneous intrusions. One set of faults may consistently displace another and would therefore appear to be later.

Strike faults, running parallel to the bedding strike, may be difficult to detect on the ground because they may not cause an obvious displacement of the outcrop. When normal strike faults dip in the opposite direction to bedding they may have the effect of repeating strata at the surface, but if the fault dip is in the same direction as bedding, strata may be cut out at the surface (Figure 2.12). The opposite applies to *reverse* faults, and unfortunately, it is not always possible to distinguish between steep normal and reverse faults on a map.

The effects of oblique or *dip faults* (where the trend is at a high angle to the strike of the bedding) are more straightforward, except that difficulties arise when attempting to differentiate between dip-slip faults and wrench faults. It is

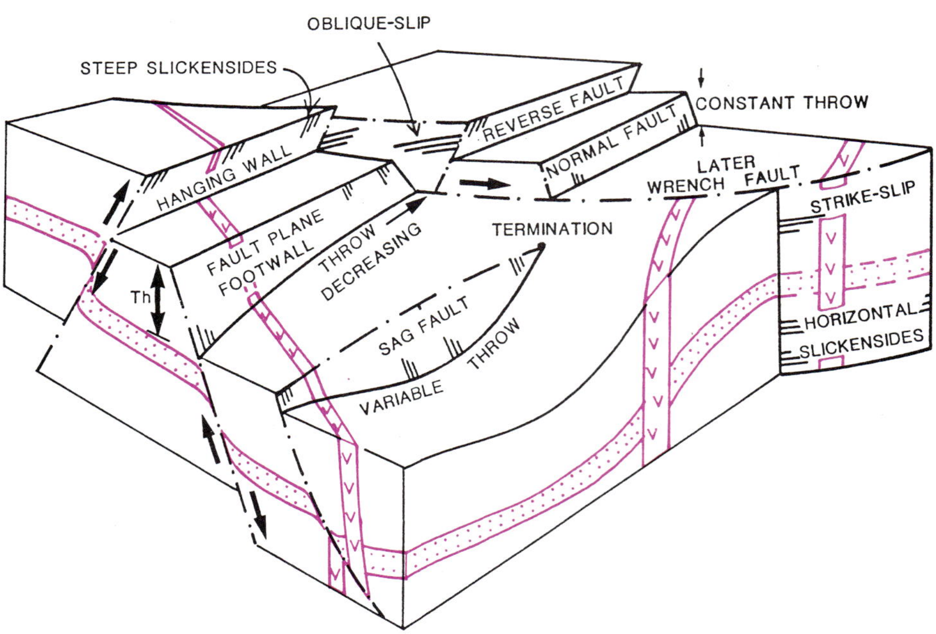

Figure 2.13 Block diagrams to illustrate various forms of slip on faults.

easy with vertical or folded strata since any sideways displacement indicates a wrench component which may be measured directly. It follows therefore that vertical dykes also give the key to fault type when dealing with inclined strata and dip faults (Figure 2.13).

To arrive at some estimate of the downthrow direction on a dip-slip fault, there are several methods to follow. Firstly, the *younger* beds outcrop on the *downthrow* side. Another method is to look down the map in the direction of dip. The bed which is displaced back towards the observer is then on the downthrow side of a dip or oblique fault (Figure 2.14*a*).

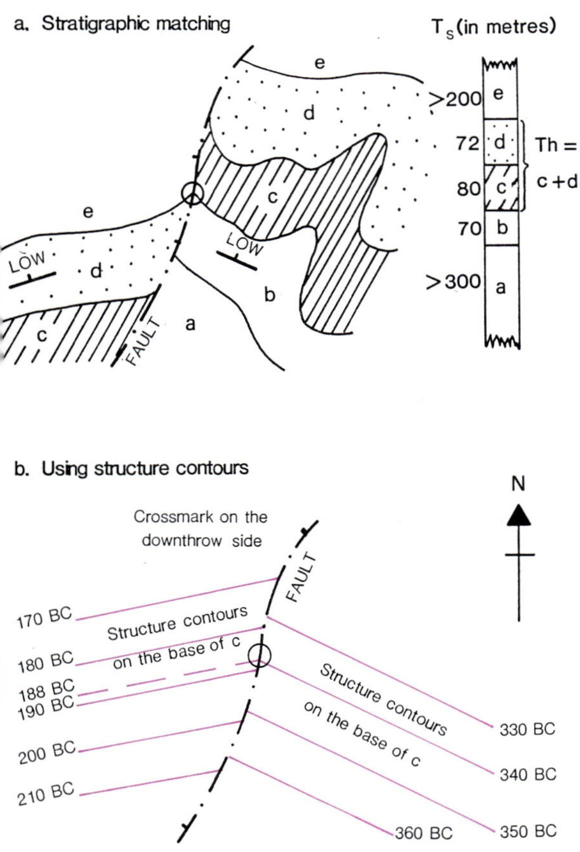

Figure 2.14 Two methods of estimating fault throw from a geological map. (*a*) Stratigraphic matching: At the circled point on fault, Throw Th $= T_{s\,\mathrm{missing}} = T_{s_c} + T_{s_d} = 152$ m to the west, and the base of c coincides with the base of e across the fault. (*b*) Using structure contours: At the circled point on the fault, Throw Th $= 340 - 188$ m $= 152$ m to the west.

The apparent vertical displacement (Th) on a published map may be obtained by using a scaled stratigraphic column, if provided at the side of the map. Matching up one horizon with another across a fault may allow the observer to make a reasonable estimate of the thickness of strata missing at the fault from the column (see Figure 2.14*a*). If the stratigraphical column is not to scale, the thickness of each missing unit must be calculated using equation (4) or (5). This gives the throw at the point of measurement, but two restrictions should be noted:

(1) If the beds are steeply dipping, the stratigraphical thickness (T_s) missing, has to be converted to vertical thickness (T_v) missing, in order to estimate the throw (Th).
(2) The throw may vary along the fault considerably. This is particularly true of scissor (hinge) or sag faults where throw may decrease to the point where the fault dies out and can no longer be traced on the map (Figure 2.13).

Fault data from structure contours. When structure contours on dipping strata can be established on both sides of a fault, for instance where boreholes are put down, then matching actual structure contour values for the same horizon across the fault gives reasonably accurate vertical displacements. These downthrow values should give actual fault plane movements for dip-slip faults but only apparent vertical displacements for oblique-slip or strike-slip faults. In engineering, the difference is of little consequence since the effect of both over a small area appears to be very similar for moderately dipping strata, although very different for vertical or horizontal strata.

This method of finding the amount of downthrow works most effectively for steep faults, and may be applied equally well to either strike or dip faults (Figure 2.14*b*) (try problem maps A.4 and A.5). However, low angle faults may pose more difficult problems which may require the additional combination of fault plane contours and structure contours.

In engineering geology, the characteristics of fault planes are just as important as their large scale geometry. Unfortunately very little information on the former can be depicted on maps except when fault planes are observed and recorded on large-scale 1:10 000 maps or more detailed site plans.

2.3.3 *Unconformable strata*

The geological processes leading to the formation of a surface of unconformity are normally well known by students of geology who study numerous examples worldwide. To the civil engineer, it is a structure only occasionally encountered on site, except in one important and common occurrence, *rockhead*. Drift deposits are almost inevitably unconformable on rockhead, since the latter often represents a major time gap involving at least several million years of erosion. There are rare exceptions, where for instance youngest Tertiary beds are followed conformably by Quaternary gravels or lake deposits. More commonly, in tropical countries residual soil has evolved from the deep weathering of rock, with which it is, therefore, conformable. In many temperate countries drift deposits are unconformable on bedrock at the rockhead surface. Similar events in the geological past involving earth movements and erosion followed by deposition, have formed the more familiar angular unconformity, and it is important for engineers as well as geologists to be able to recognise these on maps, so that their interpretation can be fully appreciated.

Recognition, at the outset, requires an understanding that the younger beds normally cut across or overstep older strata. At each point of overstep on the map, the older strata disappear under the younger unconformable boundary. In so doing they form T-shaped junctions (see Figure 2.15) which allow the unconformity to be easily followed across the map.

Because of the angular difference which often accompanies an unconformity, a sudden change in V-shape in a valley may give a good indication of its presence, especially when seen in conjunction with T-junctions (try map A.6).

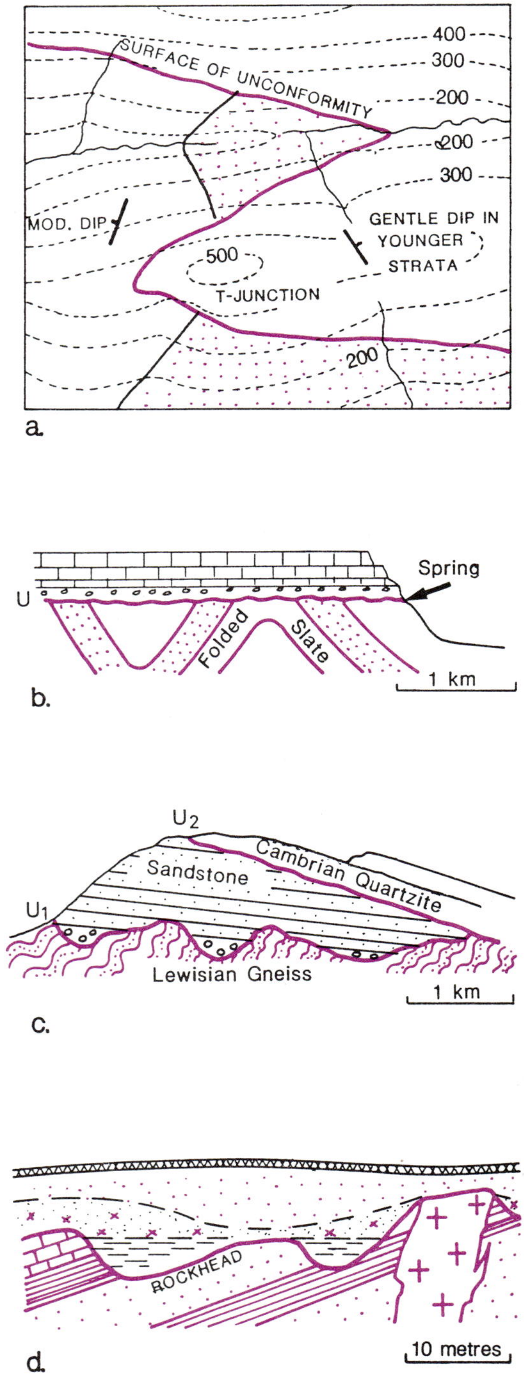

Figure 2.15 Unconformities: (*a*) Map of tilted unconformity showing T-junctions, change of V-shape and change of strike; (*b*) Section through an angular unconformity; (*c*) More complex, irregular and tilted unconformities; (*d*) Soils profile with rockhead unconformity typical of glaciated lowlands.

Unfortunately, geological maps rarely have a different symbol for unconformable boundaries and it follows that non-angular unconformities may be more difficult to detect.

Further qualitative evidence lies in the relative ages of strata involved. The younger beds always overlie the older strata (unless inversion due to tectonic overturning has taken place), and the stratigraphical gap may involve millions of years of non-deposition. One of the finest examples of angular unconformity occurs in northern England where almost horizontal whitish-grey carboniferous limestone overlies strongly folded Lower Palaeozoic greywackes and shales. All rocks of the Devonian Period are missing, so that the gap here must be at least 50 million years (Figure 2.15*b*). Prediction of unconformable boundaries in unexposed ground is possible since structure contours can be constructed on them over small areas. However, many unconformities are highly irregular, not unlike rockhead. This is why the term 'plane of unconformity' is usually avoided by geologists, especially those who have seen the relief displayed by the surface on which the Torridonian sandstones unconformably overlie Lewisian gneiss in the North-west Highlands of Scotland (Figure 2.15*c*). These conditions, as with rockhead (Figure 2.15*d*), require very closely spaced boreholes to enable the unconformity surface to be contoured.

2.3.4 *Igneous rocks on geological maps*

Generally speaking, basic igneous *lavas* which originally had a low viscosity will behave fairly predictably in forming almost slab-like flows which flood over large areas. Many of these plateau basalts, act as conformable strata and may be included in the stratigraphical column. Others form lenses, wedges or die out suddenly against former fault scarps. Some maps of volcanically active areas may be very informative and include further details of recent landslides and areas of high risk. This is well illustrated by the map of Hawaii by United States Geological Survey (Figure 2.16*a*). Interbedded *ashfalls* may also be regarded as stratigraphical in character. The more viscous silica-rich lavas generally form more limited flows and consequently are less widespread and of more complex shape.

Igneous *intrusions* display a fascinating variety of structural and geomorphological shapes due to the large number of forms and a variety of compositions which may cause variable susceptibilities to weathering.

Igneous *sills* have thicknesses from a few millimetres to many metres but are generally easy to detect since, being concordant with bedding, they are normally encountered by vertical boreholes. It is sometimes difficult to differentiate them from lava flows, except when thermal metamorphism affects the rocks above as well as below the igneous material (Figure 2.16*b*). On the map they may transgress from one stratigraphic horizon to another (see Plate 1*b*). In contrast, cross-cutting walls or *dykes* of igneous material, though easy to recognise when exposed at the surface, are often missed in drilling. Not that all dykes are vertical; in fact the student textbook view that they are straight, parallel-sided vertical intrusions may be far from the actual situation. Not only can they have a low-dipping attitude but like sills they may step from bedding to joint or fault plane and thin or thicken dramatically (Figure 2.17).

Although intrusions are commonly resistant to erosion and cause rockhead to rise, this is not always the case. Some, weakened by weathering, shearing and mineral alteration, form zones of weakness which may be more of a hazard to engineering works than most fault zones.

Sometimes more complicated igneous intrusions may be recognised on maps and sections (Figure 2.17). For example thickening and wedging of sill-like bodies at several horizons may form a *laccolith* or where sagging of strata occurs, a *lopolith* may result. Sometimes feeder dykes are recognisable, but the most common type of feeder seen on maps is the small cylindrical volcanic plug or *vent*. Occasionally they form definite hills such as Edinburgh Castle rock and the less well known Dumgoyne hill in the West of Scotland. They do not always form prominent surface features and when covered in drift may be difficult to detect (see Figure 3.14*a*).

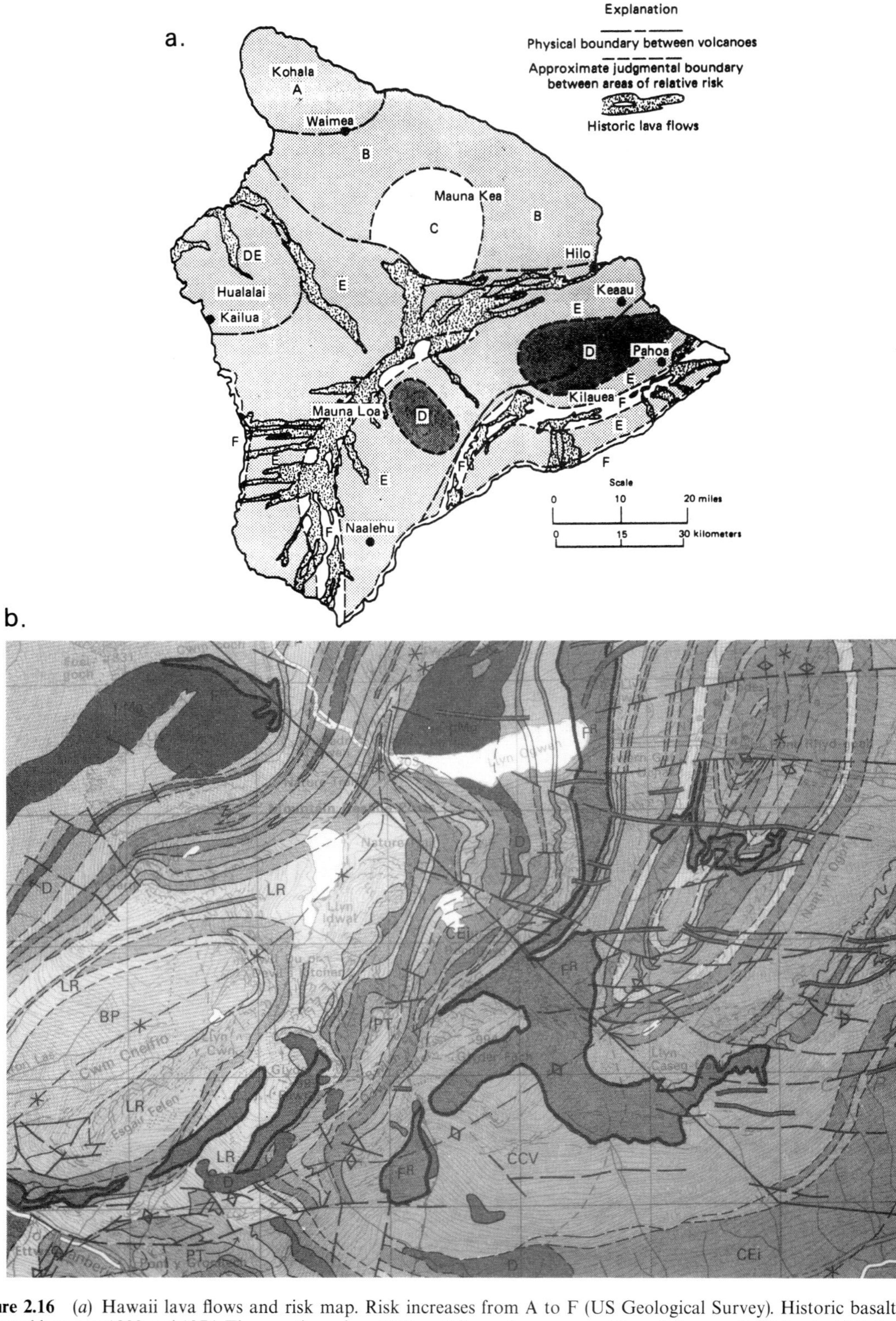

Figure 2.16 (*a*) Hawaii lava flows and risk map. Risk increases from A to F (US Geological Survey). Historic basalt flows occurred between 1800 and 1974. The eruptions since 1983 on Kilauea have occurred largely in zone F. (*b*) Ancient (Ordovician) irregular rhyolitic sills, F^R, have been outlined in bold on this monochrome extract from the BGS Bangor Sheet 106 Solid, North Wales. Scale 1:50 000. Original map in colour. (Reproduced by permission of the Director, British Geological Survey. Brown/NERC copyright reserved.)

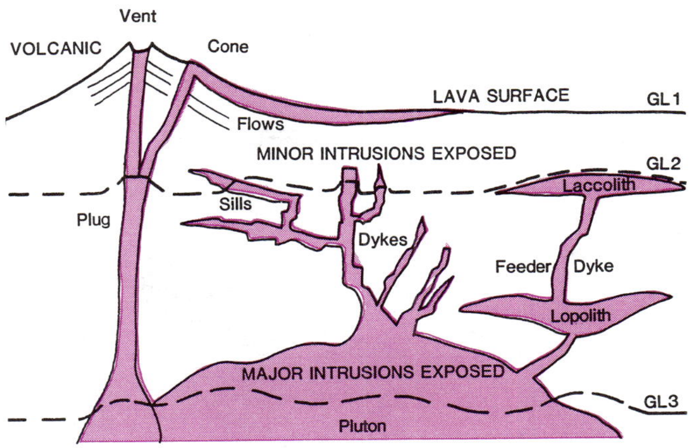

Figure 2.17 Forms of igneous bodies and their surface expression at three levels of erosion, GL1–GL3.

Dolerite vents planed off below a thin soil cover may be found by using a proton magnetometer, followed up by rotary coring. This is true of most basic igneous intrusions covered by thin soils.

Major intrusions cover large areas of a geological map. Their margins may be well-defined vertical or near vertical walls or they may have poorly defined contacts which permeate as sheets or *pegmatitic* injections cut into the surrounding rocks leaving rafts of country rock or lenses of igneous rock along a wide zone. The latter type of margin can be mapped in well-exposed areas, but may be masked by drift and very difficult to predict.

Fairly regular small *plutons* or *stocks* have narrow thermal contacts, whereas the larger plutons or *batholiths* may bake a wide zone of country rock up to several kilometres away from the intrusion. This *metamorphic aureole* is often recrystallised in such a way as to form a more resistant zone around the intrusion, resulting in difficult drilling, but this depends so much on mineralogy, alteration and weathering processes, that each case has to be judged by field inspection. In Scotland there is a marked contrast between some of the Late Caledonian forceful granitic plutons and the more complex intrusions involving migmatites (such as the Rogart Complex near Lairg (Sheet 86 S)). One of the most complex geological maps ever produced was that of the Isle of Mull (Sheet 44 D), a Tertiary igneous complex involving most types of igneous rocks, including ring dykes (Figure 2.18).

2.4 Reading geological history

The procedure for the interpretation of any geological map can follow broad lines of approach, but detailed considerations may depend on complexity of geology, map scale and quality of information available. A suggested procedure for reading all geological maps is as follows.

2.4.1 *Peripheral data*

(1) Check *scale*, *type* and *date* of map.
(2) Check distribution of *topographic* features and *date* of base topographic map, if available.
(3) Starting with the explanatory column, note type and sequence of *drift deposits* (if shown).
(4) Note how *detailed* the rock classification system is and whether thicknesses of formations or units are shown in the column.
(5) Note the *range* of ages of rock and look for gaps in the column which may suggest *unconformity*.
(6) Look for the presence of *igneous* rocks and especially minor igneous intrusions.
Metamorphic rock sequences may also form part of the map.

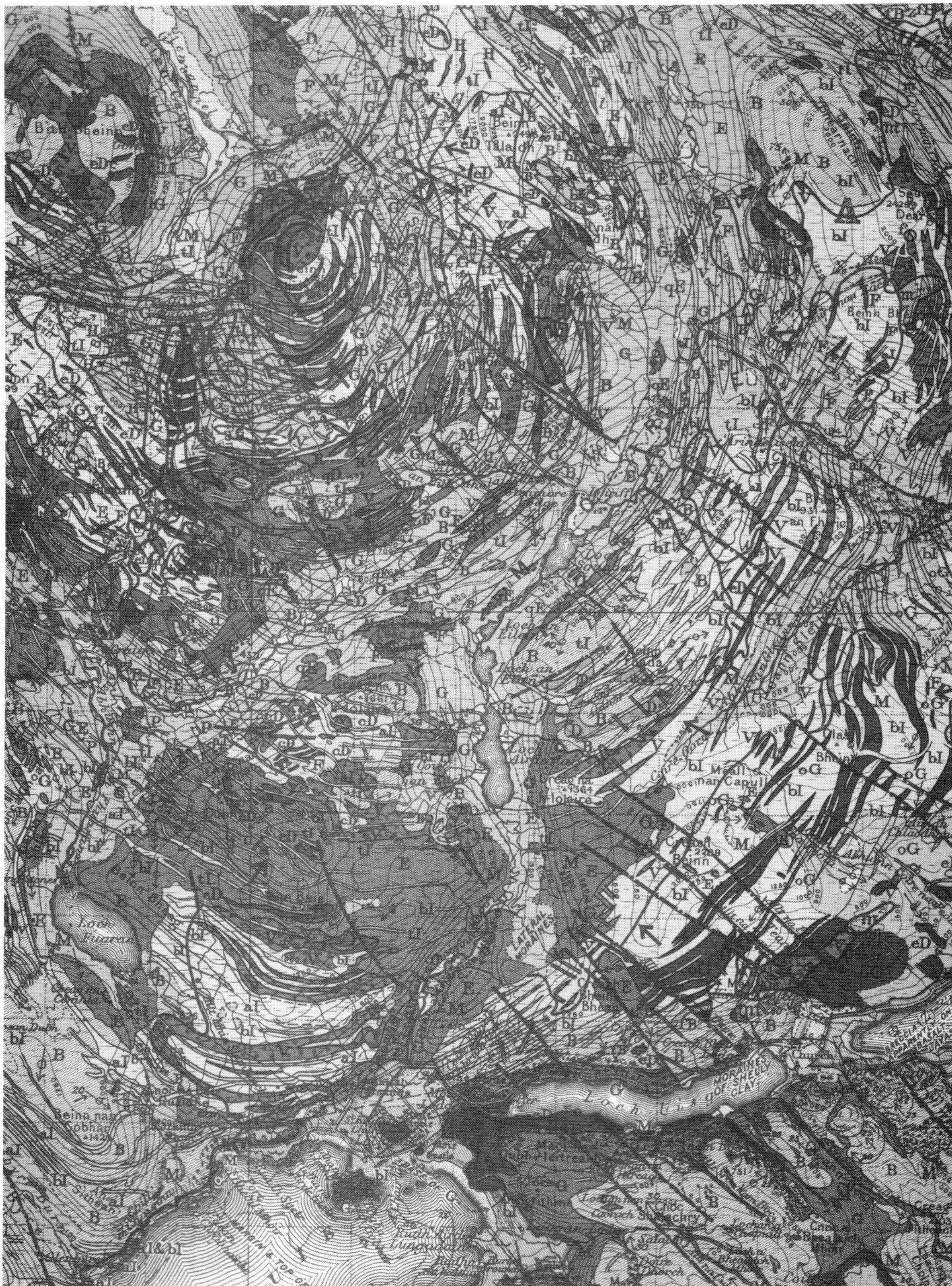
LATERAL MORAINES
MORAINES OF SHELLY CLAY

(7) Study the *key to symbols* so that when the map is scanned, all symbols will be readily understood.
(8) If a representative *section* is provided, check it for complexity.

2.4.2 *On the map*

(1) Start with the oldest rock and work up the stratigraphic column, studying the *distribution* of rocks and *dip symbols* across the area. If no dips are provided use '*the V-shape rule*' relative to topography, but note that dip symbols are useful since they also indicate presence of rock exposure.
(2) Look for the evidence of *folding* and establish fold axes.
(3) Look for the *unconformity* distribution across the area.
(4) Study the *relationship* between any faults, folds, igneous intrusions and unconformities to work out the sequence of *tectonic* events related to sedimentation.
(5) Gather as much information as possible about *lithologies* (this may involve reading a Sheet Memoir explanation or other reference text).
(6) Look at the distribution of *soils* and try to work out variations in soil sequence from one place to another (more details on this are given in Chapter 4).

After making notes and a sketch map summary, a report can be written about the geological history of the area. The way this report is constructed will depend on purpose and readership. An engineering geological history may be written with a local emphasis compared to a general geological history of the area, but in essence the data will be similarly expressed. It should deal with geological events in sequence and may have an outline as follows.

(1) Early sedimentary events including the environments of deposition and resulting lithological types.
(2) Subsequent igneous and/or tectonic events.
(3) Unconformities and younger sedimentary rocks (if present).
(4) Further (late) igneous and tectonic events (if present) and any later (post-deformational) sedimentary rocks.
(5) Soil types and their distribution. Landforms with comments on the likelihood of weathering, subsidence or landslides.
(6) Evidence of mining or former quarrying where subsidence or landfill may exist.
(7) General account of possible aquifers and any likely artesian water conditions, or other groundwater problems.

In desk study reports prior to site investigation, problems specific to the proposed site may be outlined to emphasise the need for further exploration using boreholes, trial pits and/or geophysical methods (try Questions (1) and (2) and Problem Map A.7 in Appendix A).

Figure 2.18 An extract from one of the most complex geological maps ever published. (BGS Scotland 1″ Sheet 44 Drift of the Isle of Mull, Argyllshire.) It is presented here in monochrome simply to demonstrate how it is possible to show solid geology and structure, ring and radial dyke locations, glacial data, and soil types. Even topographic contours are readable on the coloured map. The numerous letter codes and symbols are fully explained in the margins of the published sheet. Fortunately, few engineers would be expected to be able to interpret the geology of a site with such geological complexity and would normally rely on the services of a geologist for interpretation. (Reproduced by permission of the Director, British Geological Survey. Crown/NERC copyright reserved.)

3 Geological sections

3.1 Introduction

All civil engineering earthworks involve cutting through the ground surface and removing material which may then be re-used as fill or have to be dumped off site. It follows that sections through rock and soil profiles at a proposed site are important to engineers, especially when attempting to assess the quality and quantity of various materials.

Exploration geologists may also be involved in the prediction of underground structures but sometimes to depths well in excess of those affecting engineering works. Mines to 2000 m are not unknown and oil wells to 5000 m quite commonplace. In order to predict structure and lithologies to these depths and beyond, sophisticated geophysical methods have been developed which can now supply information about the lower parts of the earth's crust.

Sections can provide that other dimension which is not accurately conveyed by simply reading the map or plan. When published with a geological map, sections often follow a line along the direction of dip, which is therefore most representative of the actual geometry. Unfortunately, most tunnels and road cuttings do not follow true dip directions so that predictive sections have to be constructed from calculated apparent dips.

A brief résumé of topographic section construction is given, followed by more detailed methods of constructing vertical sections of geology from maps and later by the incorporation of borehole data.

3.2 Topographic sections

Anyone attempting to plot small-scale topographic sections will realise the need for the exaggeration of the vertical scale so that topographic variation can be more easily represented. Even on large-scale civil engineering profiles, exaggeration of the vertical scale by up to ten times may be necessary to plot subtle changes of gradient and small vertical measurements which would not otherwise be visible at true scale. However gross exaggerations, greater than ten times, look out of proportion and should be avoided if at all possible.

Construction of topographic sections is a familiar process to many students of engineering and geology, but anyone experiencing difficulty should regard them as height-distance graphs on which contours may be plotted as points at appropriate heights and distances along the *y* and *x* axes, respectively. The profile formed from the locus of these points should be smooth in most cases and take into account the location of summits, rivers and lakes on the map. Where three contours or more of the same value appear consecutively on the section it is important to continue the ground profile up or down on approaching the first, but take opposite gradients on passing through subsequent contours without crossing the level of the next higher or lower contour in the intervening ground (see Figure 3.1). Students have been known to join these points with a straight line, but they really represent undulating topography, the amount of undulation depending on the contour interval in use.

3.3 The purpose of geological sections constructed from maps

In the preliminary stages of a project, civil engineers may use geological maps to visualise the three-dimensional underground structure which has been eroded down to the topographic surface. Most civil engineers planning a tunnel or road project will be greatly assisted by a geological section along the centre line at invert or formation level. For compact sites, such as factories or power stations, parallel or serial sections enable the three-dimensional structure

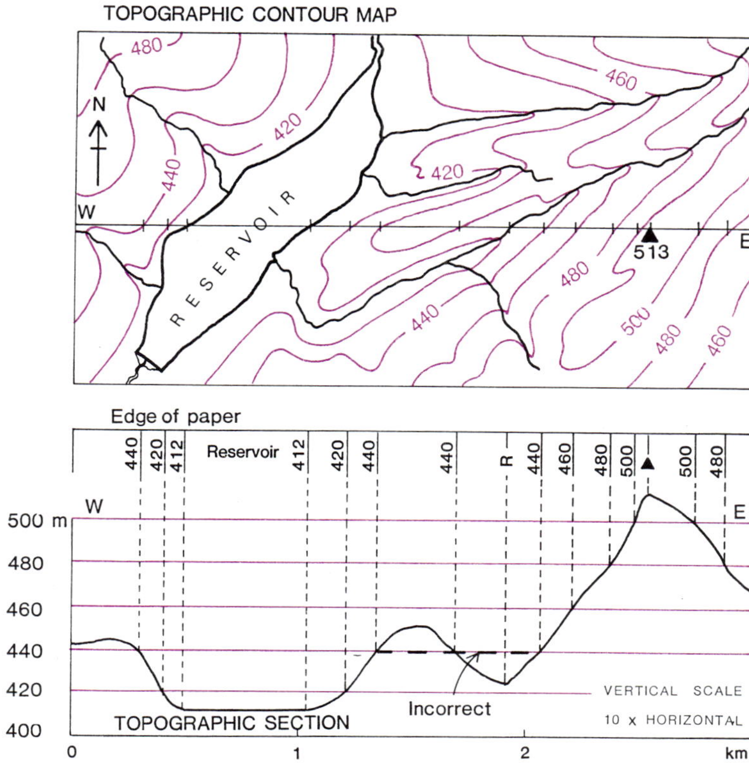

Figure 3.1 Construction of a topographic section. Note possible mis-reading of contours if all three 440 m contours are joined at that level.

to be visualised quite easily, but it is only when borehole control is added that more accurate measurement of depths and thicknesses become feasible.

3.4 Construction of geological sections from maps

3.4.1 *Accuracy*

Limitations on the accuracy of geological sections drawn without the aid of borehole control, must be appreciated both by those who construct them and those who read them. One of the greatest dangers arises when overconfident presentation provides solid-line sections, lacking any indication of the distribution of actual and extrapolated or interpolated data (Figure 3.2).

Engineers naturally require accurate sections whenever measurements are involved but this is normally only possible with an abundance of borehole data. However, at the preliminary or feasibility stages they must be prepared to accept a best estimate of underground conditions. This should normally involve the use of solid, dashed and dotted lines depending on the degree of confidence with which the section is drawn. It is also advisable to incorporate question marks wherever problems of interpretation arise.

It is sometimes wrongly assumed that near surface sections are easier to predict than deep sections. This may be far from the truth, especially when complex soils or highly variable rockhead levels occur below a site. On small scales (e.g. 1:25 000 or 1:50 000), deeper sections beyond 50 m or so are more easily constructed and less sensitive to accuracy of data, since on this scale measurements from the map cannot be made to an accuracy greater than ± 25 m. On the other hand, even where borehole data are

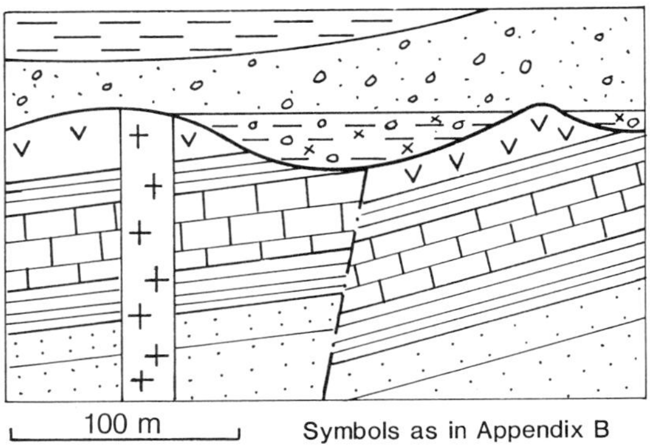

Figure 3.2 Confidence *appears* to be high on this section. But on what data is it based? A geological map, two boreholes, or ten boreholes?

available, large-scale sections greater than about 1:500 can be very difficult to construct accurately, since the limit of measurement then approaches ±250 mm on the ground, which is beyond the limits of accuracy normally expected in exploration work. Table 3.1 gives some idea of the ultimate accuracy on the ground for various scales on the assumption that it is not possible to measure a map, plan or section to a line width of less than 0.5 mm.

Table 3.1

Scale	0.5 mm line width on the ground
1:250 000	±125 m
1:50 000	±25 m
1:10 000	±5 m
1:2500	±1.25 m
1:500	±0.25 m

3.4.2 *Some basic ground rules for geological sections*

Engineers attempting a geological cross-section for the first time may require to be reminded of some geological principles which should be applied to construction, since accuracy not only depends on the data available but also the way they are used. Some fundamental points:

(1) *Sedimentary layers* are normally deposited with younger beds overlying older ones (this is known as the Law of Superposition), but it should be remembered that lateral changes in sedimentation, and therefore thickness are possible, especially when dealing with soil sections (see Figure 3.2).
(2) *Crustal deformation* may have tilted or folded rock layers to various degrees of intensity and attitude. Under conditions of brittle or semi-ductile failure they may be displaced by faulting in various directions over a range of distances. Faults cut superficial deposits only in tectonically active (earthquake) zones.
(3) *Igneous intrusions* normally cut across sedimentary or metamorphic units but could be folded if pre-deformational in age. If they are post-deformational they generally follow planes of weakness in the rocks which could include bedding planes. They cut through superficial deposits only in areas of recent volcanic activity.
(4) *Unconformities* may involve large angular changes of dip on a section. It should be remembered that, when they occur, younger strata (including deposited soils) overstep the older strata on a single unconformable and perhaps irregular surface (see Figures 2.15 and 3.2).
(5) *Superficial deposits* depicted on sections normally form discontinuous layers whose depth is difficult to establish without boreholes. They may have to be shown diagrammatically or even omitted on sections drawn from geological maps.

3.4.3 *Vertical exaggeration*

An initial assessment of ground conditions may be derived from small-scale sections at the research stage of a project. They are most beneficial for extensive tunnel or road routes prior to the specification of a site investigation programme.

It is expected that some *vertical scale exaggeration* (E) will be applied to the section for the sake of clarity. Even an E of $\times 2$ will show up a 300 m hill at 1:50 000 scale as 12 mm above sea level instead of the rather diminutive 6 mm at true scale.

Exaggeration has the effect of:

(1) increasing all dips of planes on a section;
(2) increasing the amplitude of folds;
(3) thickening strata as they approach a horizontal dip even though the stratigraphical thickness remains constant in reality.

These disadvantages are a sacrifice for the benefits of being able to include thin but significant layers of strata which could not otherwise be depicted. However, unless kept under control, increasing the vertical scale can cause distortions of a section which look unrealistic, and eventually become unplottable as most structures appear to be very steep (see Figure 3.3). On large scale sections, exaggeration may be unnecessary.

3.4.4 *Correction of dips along the section*

Whilst dips to be plotted on a geological section must be corrected for direction of the section if not parallel to the true dip in the area, they must also be corrected for any vertical exaggeration which may be applied. For example, a tunnel line cutting obliquely across a true dip $\alpha°$ at a horizontal angle $\theta°$ will produce an apparent dip $\alpha_A°$ on a true scale geological section. So that (Figure 3.4),

$$\tan \alpha_A = \tan \alpha \cos \theta \qquad (7)$$

For values of $\theta > 15°$ the correction becomes necessary. An alternative graphical solution involving a stereonet is given in Appendix C.

However, if the vertical scale exaggeration on the section is $\times E$, this must also be taken into account by finding the ultimate corrected dip α_E. Thus,

$$\tan \alpha_E = E \tan \alpha \cos \theta \qquad (8)$$

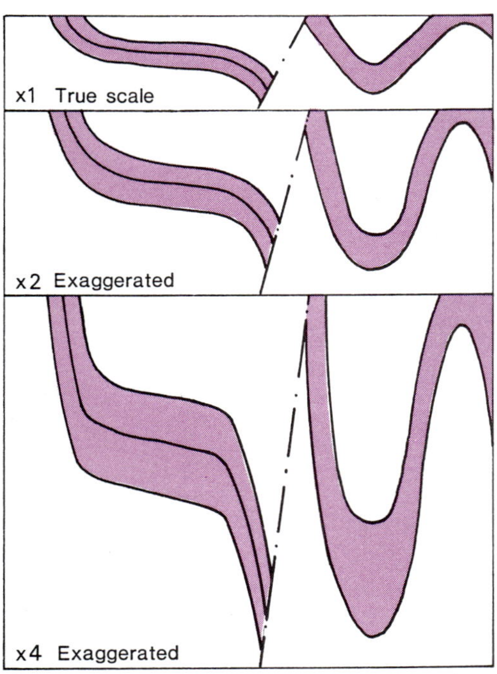

Figure 3.3 The effect of vertical exaggeration of scale on the portrayal of dips and thicknesses in section.

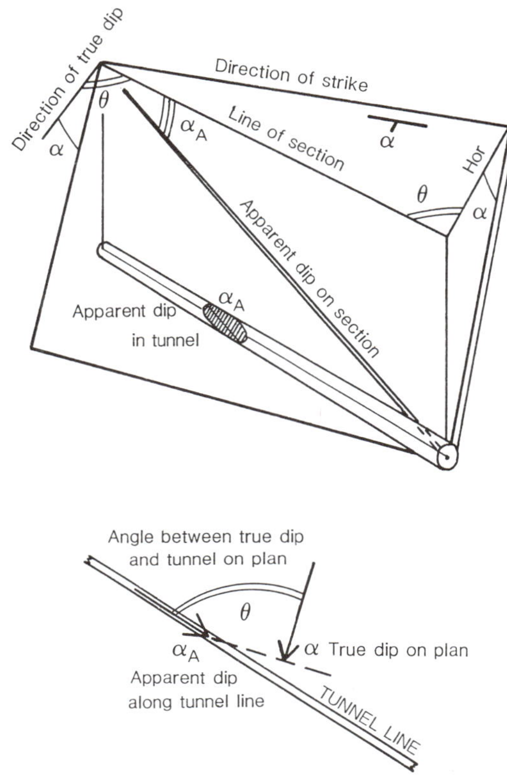

Figure 3.4 Diagrams relating true dip to corrected dip along the line of section or tunnel. Note: If $\theta < 15°$, then $\alpha \doteqdot \alpha_A$.

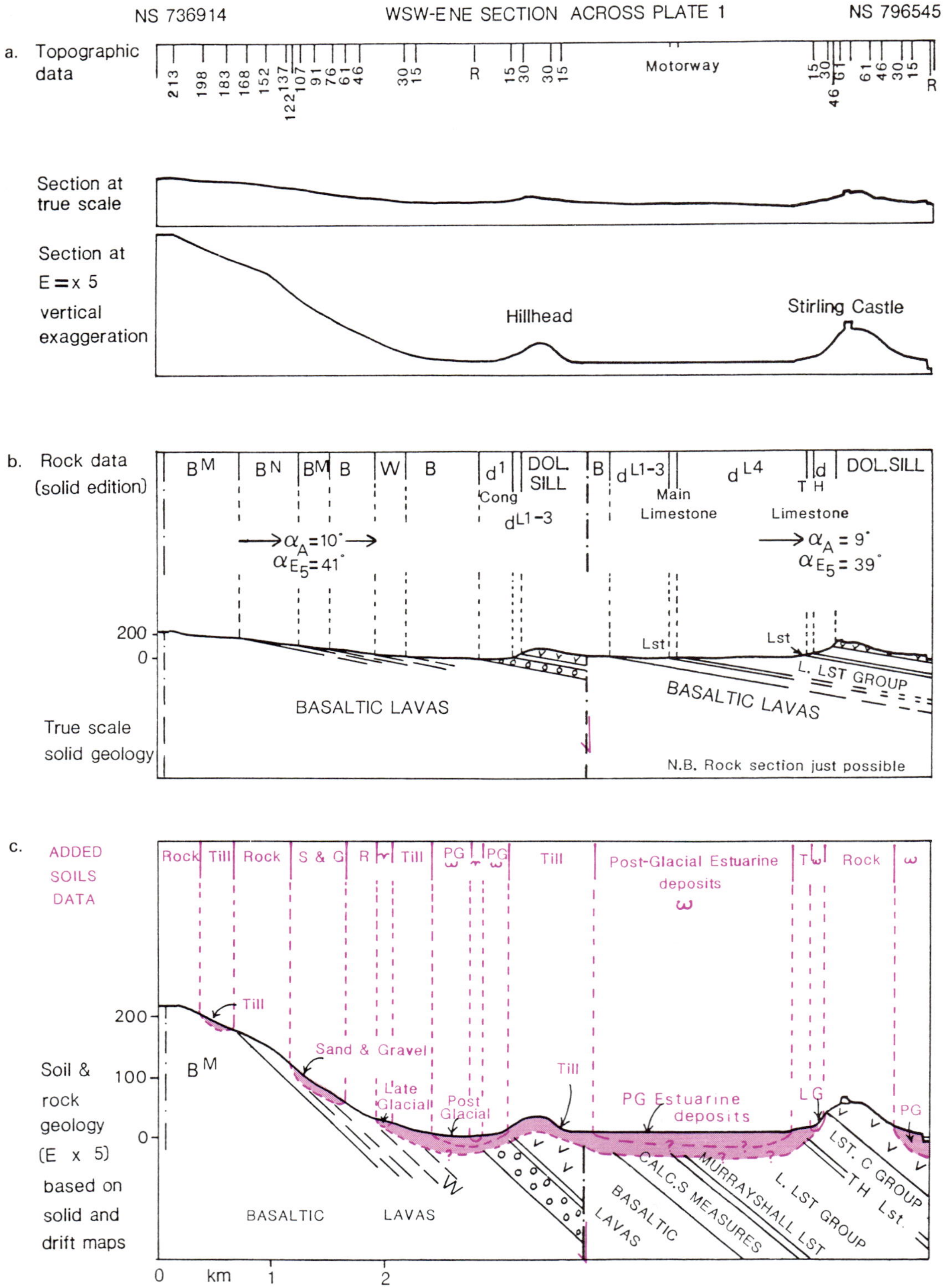

Figure 3.5 Geological section construction (for symbols see Plate 1 and explanations).

3.4.5 *Procedure*

Recording map data along the proposed line. Use a strip of paper (the *working strip*) or the edge of a sheet of paper and place it along the proposed section line on the map.

(1) Mark the paper with the end points of the section and any change of direction between those two points.
(2) Mark any stratigraphical (or soil) boundaries using a short line up to 10 mm long, perpendicular to the edge and label the intervening areas with the materials present (Figure 3.5*b*).
(3) Mark any structural features (such as faults, unconformities, fold axes or igneous intrusions) using longer lines perpendicular to the long edge of the paper approximately 20 mm in length and name any major structures.
(4) At an intermediate level (say 15 mm in from the long edge of the paper), place dip arrows representing apparent dips corrected, if necessary, for exaggeration (α_E°). These should have been derived from map dips or dips calculated close to the section line. The arrows should point either right or left along the paper with the apparent dip (α_E°) at the head of the arrow. Where dips are simple or changing systematically this process is straightforward, but if frequent changes of orientation are shown on the map, some care as to their position on the working strip has to be exercised.

Plotting the geological section (Figure 3.5*b*). Assuming the appropriate topographic section is available (or see Figure 3.5*a*), the boundaries in (2) and (3) above should initially be cross-marked in pencil in their appropriate position on the section at ground level. Unless the dip of a fault plane or dyke is known it should be plotted vertically to cut the section below rockhead level, if this is shown. Although only a first approximation, on a small scale section it is an acceptable limitation, and warns of the possible need for an angled borehole.

If the geological section involves both soils and rocks, rockhead levels have to be approximated by reference to Drift maps. Without borehole control the superficial cover is necessarily diagrammatic and rockhead may have to be represented by a uniform strip varying to a depth controlled largely by the frequency of exposures in the area. Whilst in Britain an average depth of drift may only be a few metres with a maximum little more than 40–50 m, some more active (*neotectonic*) areas may have considerably greater depths of overlying soils (e.g. the former lake deposits in the Kathmandu valley of Nepal reach a maximum depth of over 120 m before rockhead is encountered).

The drift sequence may then be inferred from its distribution over the area involved. For glacial soils, lodgement tills formed during ice advance may underlie either glacial ablation tills or late-glacial sands and gravels (see Figure 3.5*c*).

If post-glacial estuarine or alluvial deposits are present, they should occur above the glacial deposits and in turn be overlain by more recent soils such as peat or material involved in landslips.

Dashed lines and question marks may be necessary at this stage since solid lines give the impression of high confidence levels. This can be especially misleading if the sources of information on which the section is based are not indicated.

Plotting rock boundaries. Sedimentary and metamorphic rocks are plotted from surface outcrop data corrected for dip along the section. This is essential if boundaries are to be projected to any depth below ground level.

(1) Using the corrected dip nearest to the boundary to be plotted, project the boundary below ground at the appropriate corrected angle from the horizontal. Normally a straight pencil line a few centimetres in length is drawn for each known geological boundary, but faults, dykes, plugs and other major intrusions will cut across these.

(2) If the rock succession is dipping uniformly along the section, the procedure will be relatively simple (Figure 3.5*b* and *c*), but if folds are present, curved boundaries must be plotted (see Figure 4.6). If the vertical exaggeration is significant, slight wedge shapes may be unavoidable. Normally, unless known thickness changes occur on the map, or on the steep limbs of folds, it is important to curve the layers to maintain the stratum thickness (T_s). Where dips increase in angle along the section, convex curves are plotted below ground; where dips decrease, concave curves are necessary. The curvature may be estimated by using pencilled in 'normal' lines perpendicular to the dip, which help to maintain stratigraphical thicknesses. However, the curvature is more difficult to reproduce when steep limb thinning may have occurred due to ductile strains in zones of high deformation.

(3) Work out fault displacements from the map by the methods given in Section 2.3.2 and apply them using the vertical scale.

(4) Plot any unconformities by obtaining the dip of the rocks immediately above them. Remember, unconformities may also be folded.

(5) Mark up the section with names of formations, lithologies or structures where possible, so that a key becomes less essential and the section is easier to read.

(6) At depth, dashed lines and question marks again become important to emphasise the uncertainty which is inevitable without borehole control.

3.5 Construction of geological sections incorporating boreholes

Ground conditions which control the detailed design of most civil engineering works may be most accurately predicted when surface data is supplemented by borehole or trial pit logs. Boreholes should normally be taken to a depth several metres below the level at which any load influence or excavation is likely to occur (i.e. below formation level for cuttings, below invert level for tunnels or underground openings and below the depth of influence of foundation loads). If mine workings or other suspect strata are expected then the depth should be increased accordingly, to enable correlation to be achieved and more accurate sections to be constructed without the unfortunate absence of information in the *data gaps* (see Figure 3.6).

Although the value of competently logged boreholes cannot be overemphasised it should be borne in mind that cylindrical cores only a few centimetres in diameter are unlikely to be fully representative of the three-dimensional rock mass and can easily miss vertical structures such as dykes, faults or joint sets (see Figure 3.7). Inclined drilling may be necessary to reveal more

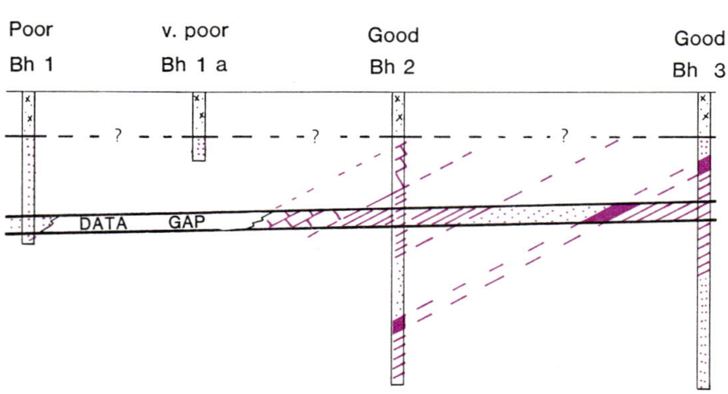

Figure 3.6 Simplified geological section showing the value of deeper cored boreholes in predicting lithologies and reducing data gaps.

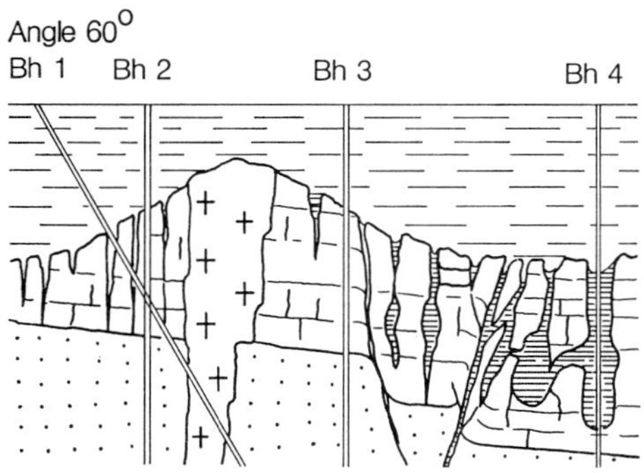

Figure 3.7 Sections from boreholes. Boreholes 2 and 3 give a similar, but false impression of horizontal rockhead, strata and joints. Borehole 4 gives a low rockhead level, but Borehole 1 picks up most structures. Inclined boreholes should be considered for occasional use where appropriate.

accurately a situation where vertical structures are dominant, but care must be taken when plotting borehole inclinations on vertically exaggerated sections.

Using the exaggerated section for vertical or horizontal measurement works well, but no angular measurement, either in degrees of inclination, or in length, may be attempted without corrections. Angles should be replaced by gradients and the appropriate geometry or trigonometry adopted (Figure 3.8).

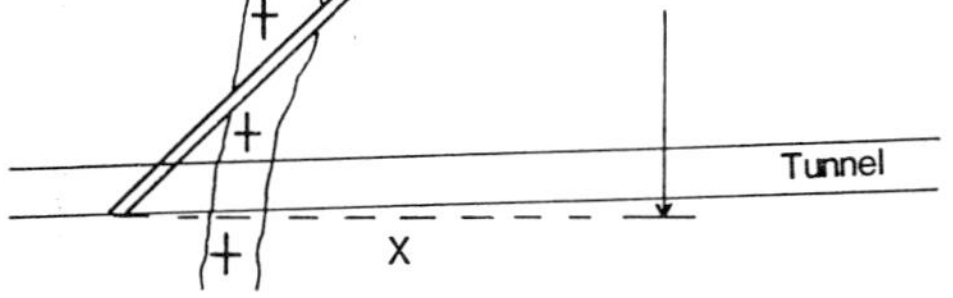

Figure 3.8 An exaggerated section illustrating the effect of increasing vertical scale on angular measurement. l = drill length to invert, x = horizontal distance, d_E = depth to end of drill length allowing for tunnel gradient, and γ_E = exaggerated angle on section. Whilst x and d_E can be directly scaled from this section, neither γ_E nor l should be measured directly. See text for formulae.

The length (l) of an inclined borehole must be calculated from an exaggerated section since l cannot be measured directly, but both the horizontal distance along a tunnel (x) and exaggerated depth (d_E) (which includes any gradient effects in the tunnel) can be measured. Hence

$$l = (x^2 + {d_E}^2)^{\frac{1}{2}} \qquad (9)$$

Also the actual rake (γ) can be calculated since

$$\tan \gamma_E = E \tan \gamma \qquad (10)$$

This of course assumes that the rig is situated on the ground surface directly above the tunnel line.

True scale sections allow both angular and linear measurements to be made providing they also lie within the plane of section.

On a large scale, the correction formula (8) for apparent dips becomes more important since even small changes of dip show up significantly and must accommodate borehole data taken on the line of section. Ignoring dip and other measurements made on exposures at the surface may have serious consequences for the interpretation of rock structure at depth (see Figure 4.5).

Even where surface exposure is lacking along the section line, an understanding of the local and/or regional geology from maps and memoirs should assist the construction of more valid geological sections. Surface exposures become even more valuable in mountainous country where deep boreholes would be extremely expensive (see Figure 4.6).

3.6 Correlation of sub-surface layers

Most ground investigations involve boring in soils down to rockhead level. Whenever the soil distribution across the site is of importance, a geological section using borehole logs should be attempted although great care has to be exercised by engineers when quantities or measurements are to be taken.

3.6.1 *Correlation of superficial deposits*

The varying modes of formation of soils make

their three-dimensional distribution difficult to predict. Local changes in the balance between erosion and deposition, especially when associated with glaciation, often lead to chaotic distributions and sometimes make correlation impracticable. The simplest solution is to join lithological changes in adjacent boreholes by straight lines, but since this is almost always inaccurate, dashed lines with accompanying question marks must be used.

Multiple glacial advances and retreats will complicate a sequence. Glacial advance of ice sheets involves erosion of bedrock and deposition of lodgement or sub-glacial meltout tills. Retreat or standstill is more complex with ablation and fluvio-glacial deposits often juxtaposed or interbedded. Most estuarine, river or lake deposits are not only post-glacial but may have a more regular and hence predictable, cross-section (e.g. Grangemouth on the Forth estuary, Figure 3.9).[2]

3.6.2 *Correlation of rockhead*

Rockhead is the most frequently misread surface below ground. It is a very signficant level for both excavation and foundations. Its condition and elevation depend on the more recent geological history of an area and this in turn is largely dependent on climatic conditions over the last few million years.

Recent deposits overlying the rock may result from wind, river or glacial processes or perhaps a long period of active weathering. Residual clays are common in tropical climates since they result from deep chemical decomposition of rock so that fresh rockhead may eventually be encountered several tens of metres below the

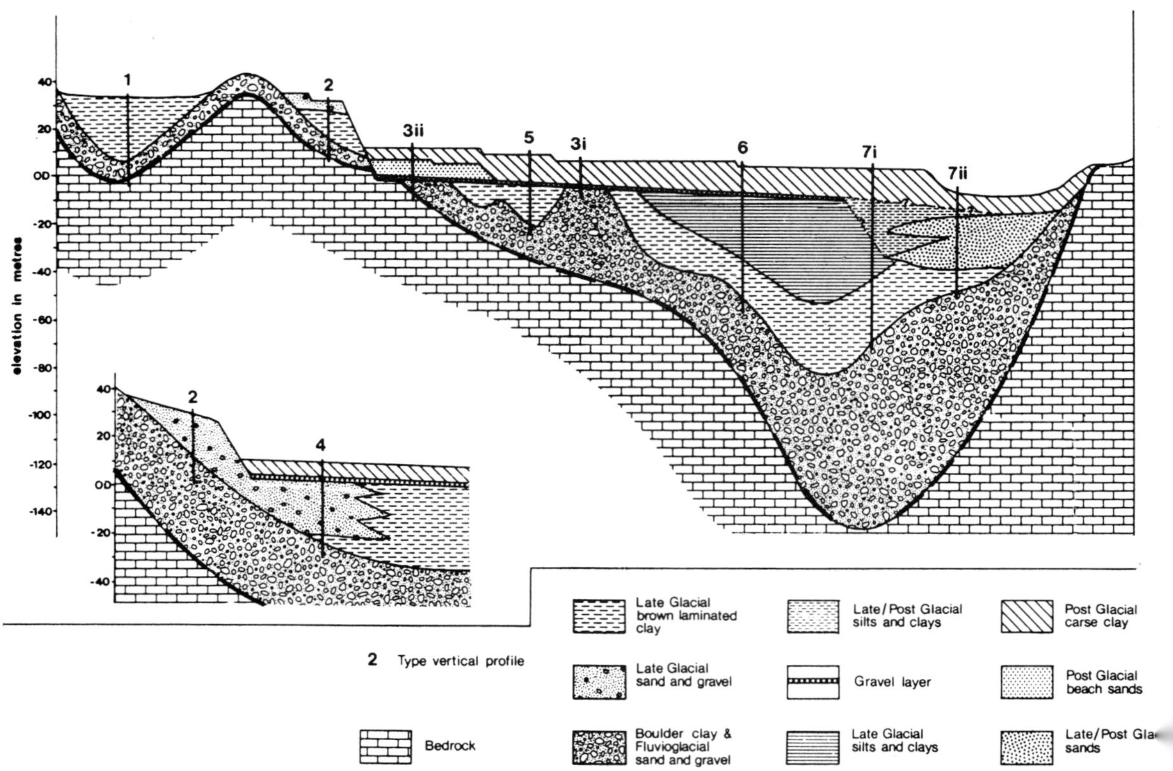

Figure 3.9 Geological sections of the Grangemouth area on the upper Forth estuary, showing how 'profile types' numbered 1–7 were obtained by Gostelow and Browne, 1986.[2] (Reproduced by permission of the Director, British Geological Survey. Crown copyright reserved.)

surface. In temperate climates, some rock masses may not have experienced recent erosive processes and the chemical effects may only reach a few metres below the surface, with the changes from fresh rock down to residual clay being irregular and difficult to pinpoint.

Marine and mature-river planation of rock may result in a predictable and fairly level rock surface below the later deposits, but where erosion of channels or wind deflation has been involved, more surface irregularity may be expected.

One of the most unpredictable processes affecting rockhead is direct glaciation and its associated sea-level changes. Glacier ice scours and plucks the rock over which it moves, taking advantage of zones of weakness and leaving more resistant rock masses upstanding in its path. In lowland areas, this effect decreases as the ice sheet spreads and its erosive power diminishes, but nearer to the mountain glaciers rockhead may be extremely irregular, especially where lithological contrasts exist (see Figure 2.15*d*).

A further problem associated with correlation of rockhead is the buried channel where entrenchment has occurred during a sea level drop, leaving a deep gorge later to be filled with

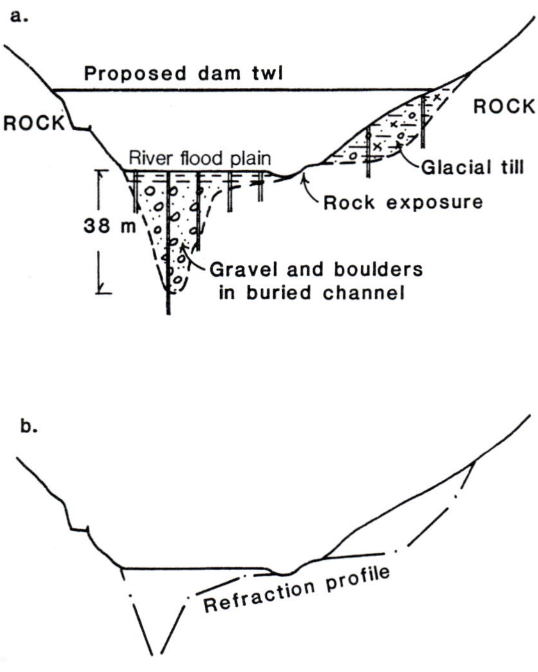

Figure 3.10 (*a*) A sketch cross-section of a glaciated valley where rockhead was revealed by preliminary seismic survey and (*b*) confirmed by drilling. The dam site was not chosen because of the buried channel and weak abutment.

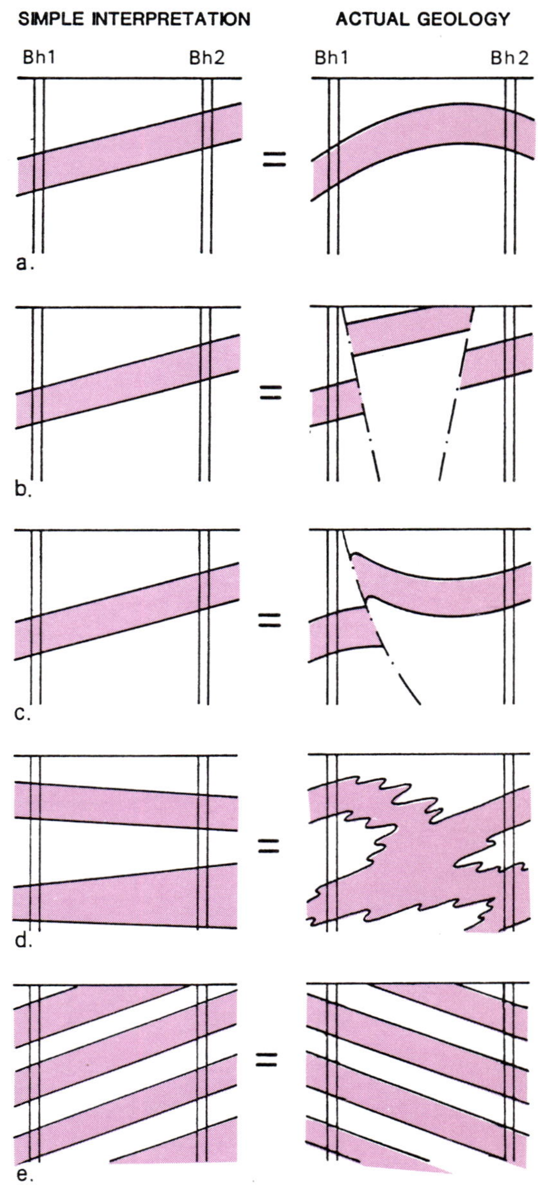

Figure 3.11 Comparison of simple and actual correlations of rock between two boreholes.

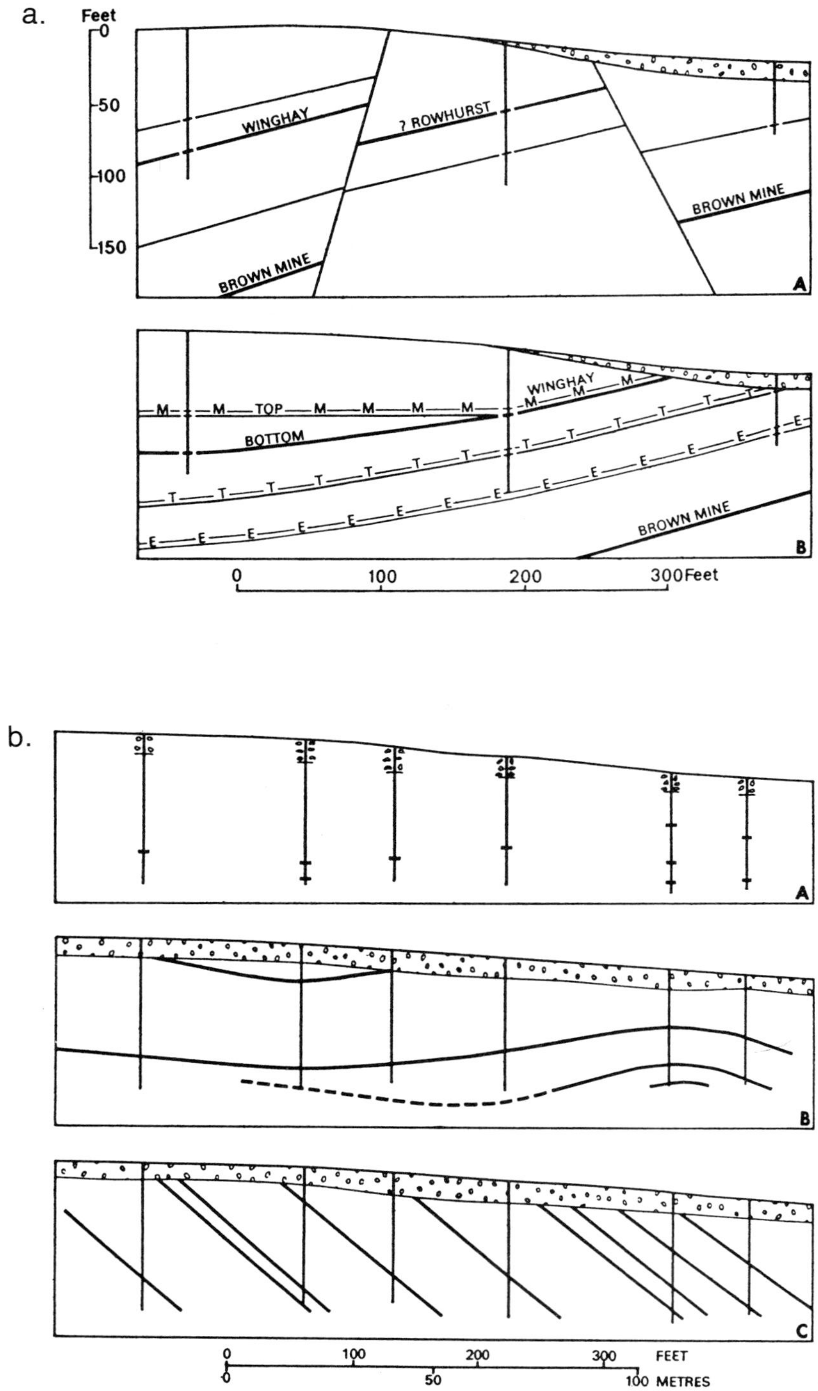

Figure 3.12 (*a*) Interpretation of three boreholes in Potteries Coalfield: A, based on driller's logs only; B, on detailed examination of cores. (*b*) Site investigation on South Crop of South Wales Coalfield: A, information plot of six water-flush holes; B, 'obvious' interpretation; C, interpretation based on local knowledge of sequence and structure. (From Woodland, 1968.[25])

coarse clastic alluvial deposits as the sea level rises again. The present river course may bear little relation to the earlier channel (Figure 3.10*a*). Where dams, tunnels or other major excavations are involved rockhead level can be critical, and where appropriate, seismic surveys may supplement boreholes in assisting the construction of accurate rockhead profiles on geological sections (Figure 3.10*b*).

3.6.3 *Correlation of rock*

In comparison to soil prediction and rockhead profiles, the drawing of sections to include rock lithologies and structures may seem straightforward. On many sites this is certainly the case since rock layers are often constant in thickness and dip regularly over large areas. However, the assumption of a simple structure has led to major errors in rock assessment on civil engineering sites and problems have not been foreseen, due to oversimplification.

Some basic examples of misinterpretation are given in Figure 3.11. The main rock prediction problems include:

(1) missing vertical structures such as faults, dykes and vents (see Figures 3.7, 2.15*d*, 2.13);
(2) misinterpretation of folded structures;
(3) misjudging lithological characteristics such as strength, hardness, or discontinuity spacing;
(4) not fully investigating coal seams and other mineral workings;
(5) not predicting groundwater inflows.

Most of these can be overcome by better spacing and logging of boreholes combined with correct rock testing and an appreciation of the local conditions of the area being investigated (see Figure 3.12).

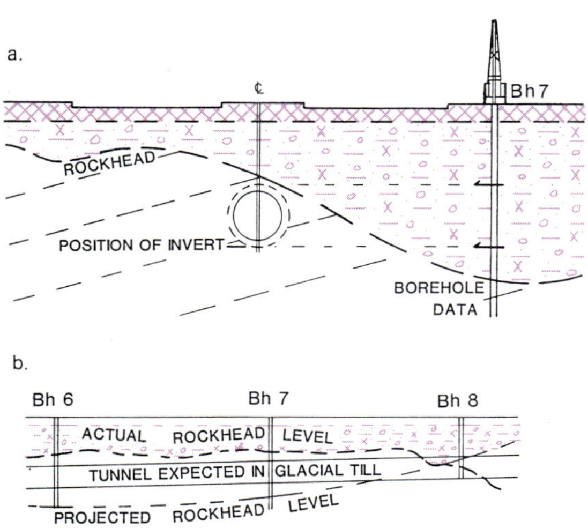

Figure 3.13 (*a*) Boreholes for a proposed sewer tunnel put down remote from the centre line. (*b*) Consequence of plotting the borehole data on to a remote longitudinal section. Note: the sketch sections are not to scale.

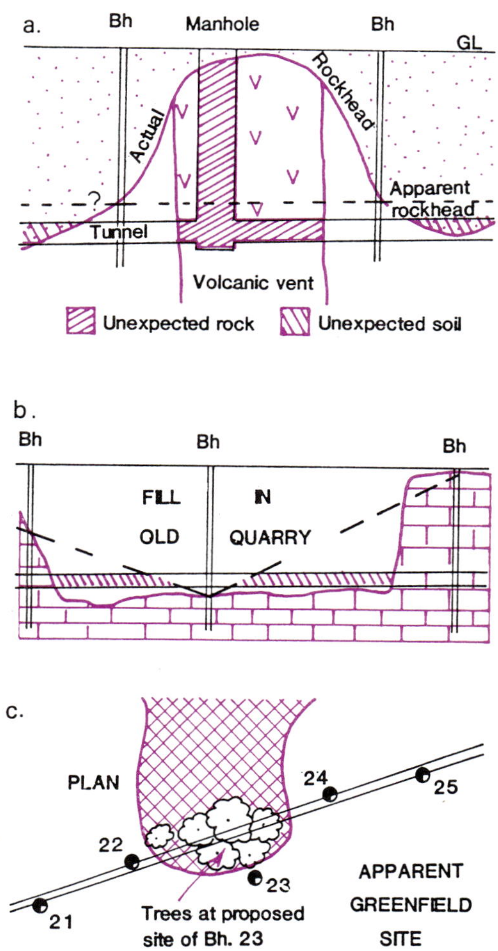

Figure 3.14 Incorrect interpolation of rockhead between boreholes: (*a*) Where a sharp rockhead rise is missed; (*b*) Where the edge of an old quarry is not defined; (*c*) Where an offset borehole detects nothing unusual.

One of the main sources of inaccuracy in geological sections is the offset positioning of boreholes which is sometimes necessary due to surface obstacles, but may be due to lack of precision by both contractor and engineer. Even a few metres off-line, what appears to be an accurate tunnel section becomes unreliable with respect to both rockhead level and lithological distribution at invert (Figure 3.13). Occasionally little can be done to predict a quirk of geology or a man-made intervention between closely spaced boreholes even when these are on line (Figure 3.14*a* and *b*).

3.7 Three-dimensional representation

There are times when it becomes appropriate to try to predict the third dimension in a more comprehensive and useful way if site conditions require it.

3.7.1 *Serial and multiple sections*

The construction of numerous parallel (or serial) sections is helped by early planning of borehole positions following the research phase of a site investigation. Serial sections are most appropriate for a large compact site such as a factory or power station since they enable a composite diagram to be drawn.

However, it may not be possible to position boreholes on a grid basis for serial section construction, but even with a more irregular pattern of boreholes it is sometimes useful to construct intersecting sections in a *herringbone*, *box*, or *ribbon* pattern. Block diagrams may be drawn from either serial or multiple sections and represented isometrically. Those illustrating only selected geotechnical features can sometimes be most effective (see Figure 3.15). Alternatively, each section may be cut out and the whole presented as a model. This helps the important visualisation process and allows more accurate measurements to be taken.

3.7.2 *Models*

For solid models to be effective they must be cut

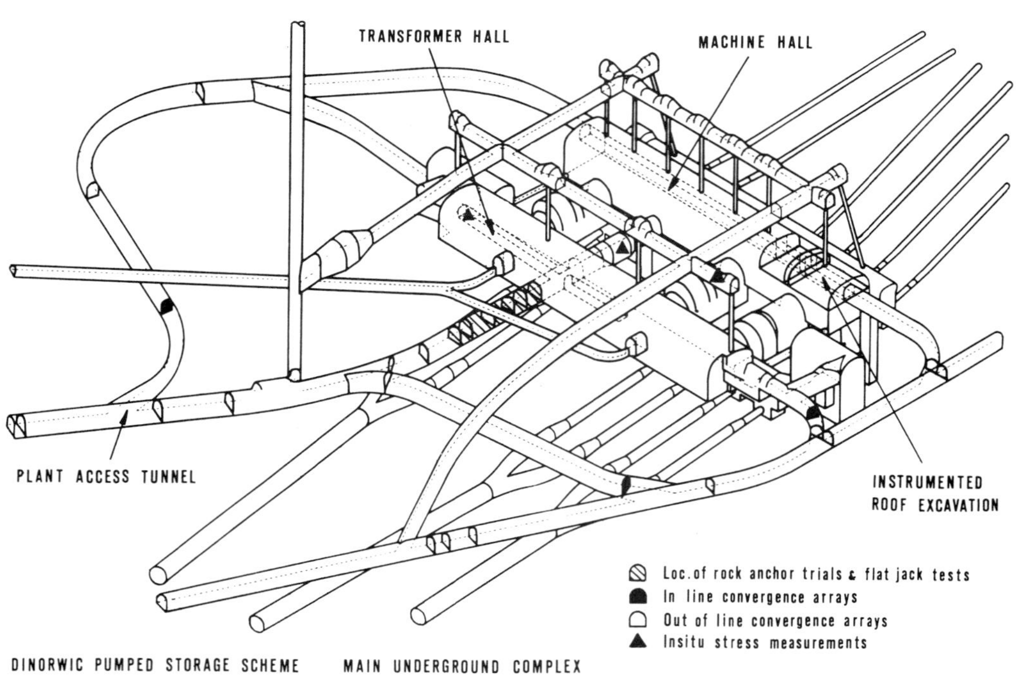

Figure 3.15 Clear indication of spacial positions of rock tests in a complex three-dimensional rock excavation by effective use of a block diagram. (Courtesy of James Williamson and Partners, Glasgow.)

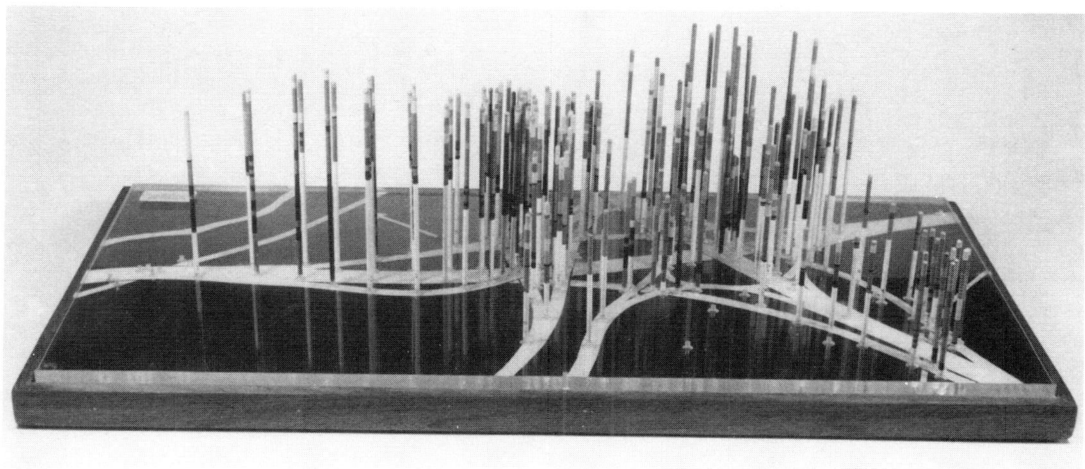

Figure 3.16 Model showing the three-dimensional relationships between boreholes at a complex motorway interchange. Any section can be obtained by placing plain paper behind the required boreholes. (Photograph courtesy of D. Wallace.)

by numerous cross-sections, and the cost of construction may not be justifiable except on large or complex sites. Perspex models are also expensive to produce but may be valuable for demonstrating complexities.

One of the most cost-effective ways of representing three-dimensional distributions from numerous boreholes is by using dowel models. These are produced by mounting the borehole site plan on 20 mm board or perspex and drilling 10 mm deep holes at each borehole position. Small diameter dowels or metal rods are then cut to length (using reduced levels from the borehole logs) and colour coded for each category of soil and rock encountered. A base datum is decided for all rods (usually below the deepest borehole) which are then fixed at that level on the surface of the board (see Figure 3.16). Once in position, a clearer three-dimensional image is available, which may be enhanced by the use of coloured thread to pick out individual horizons such as rockhead. This has been done for some large and complex highway interchanges where a comprehensive description of ground conditions would otherwise have proved very difficult.

Three-dimensional packages for microcomputers, such as Camsoft 'Surfer' are becoming available, so that in future it should be possible to feed in coordinates and levels which then allow viewing of the graphics model from any angle. Hard copies of any required section would then be made available (see also Paul and Balfour, 1990[27]).

Chapter 4 Engineering geological maps

4.1 Introduction

The earlier sections of this book have shown how valuable conventional geological maps can be if used as an integral part of any planning procedures for development schemes. They have also pinpointed some of the shortcomings of these maps when used for engineering applications. There are many ways of improving existing maps by incorporating further field data or perhaps by some re-classification of legends. This has been done on individual sites for many years and will continue to be the main source of preliminary data compiled to suit a particular site and its problems.

It is interesting to recall that geological mapping based on matching strata characterised by certain fossils (stratigraphical palaeontology) was first carried out by a civil engineer, William Smith, in the late 18th century. It was initially based on detailed observations of the strata during the construction of canals through the English countryside and later became a full time occupation as Smith[3] covered large areas of England and published numerous maps (see Figure 4.1). Although this pioneering work was later developed by geological surveys throughout the world, purpose-made engineering geological maps have only become available in the second half of this century.

Some of the earliest engineering-based geological maps were made in the USSR in the 1930s, but it was not until the 1960s that specialised geological maps began to appear worldwide. There has been some attempt to reach national and international agreement on classes and format of engineering geological maps and especially on the symbols which should be used (see Appendix B). It is generally recognised that the widely varying needs of different sites together with limitations on available data are bound to lead to individualistic approaches for some sites (see Section 4.4.2). Nevertheless, as a consequence of the proliferation of numerous styles of map, the Engineering Group of the Geological Society of London established a working party, consisting of ten experienced engineering geologists, to consider standardising various aspects of engineering geological maps and plans.

In 1972 the working party report was published in the Quarterly Journal of Engineering Geology under the title 'The Preparation of Maps and Plans in Terms of Engineering Geology'[4]. It covered in some detail possible forms of geological information and recommended symbols which could be used on both *Regional Engineering Geological Maps* (at 1:10 000 scale or smaller) and *Engineering Geological Plans* (at scales greater than 1:10 000).

A subsequent publication of Dearman and Fookes (1974)[5] continued this theme with the addition of a third category known as *Geotechnical Plans*, on which important soil or rock parameters are selectively depicted, largely as a result of field and laboratory testing. They also emphasised the three stages at which maps, plans and sections would be required in any major project. These were:

(1) the reconnaissance stage—prior to and for planning sites and site investigations;
(2) the site investigation stage—incorporating the results of the investigation;
(3) the construction stage—when remedial work could be assessed following excavation.

They also placed more emphasis on the value of geomorphological (landform) mapping for highways, in which detailed shapes and slopes of all surface features are plotted on site plans.

In the 1976 publication of *A guide to the presentation of Engineering Geological Maps* by UNESCO,[6] the treatment was broadened to take into account the requirements of countries much larger than Britain, but often less well endowed with geological maps. The book was written by the Commission on Engineering Geological Maps set up by the International Association of Engineering Geology (IAEG).

The Commission concluded that maps may be

Figure 4.1 Monochrome copy of part of William Smith's 1815 geological map of England showing the distribution of Mesozoic formations in Wiltshire.

classified as small, medium or large-scale and be *multi-purpose* (covering many aspects of engineering geology) or *special purpose* (concentrating on one specific aspect). They further recognised the need for either *comprehensive content* (where zoning or grouping occurs) or *analytical content* (where detailed variations are depicted).

These criteria covered numerous combinations of maps from small-scale regional reconnaissance, based on photogeology and terrain evaluation in developing countries, to large-scale geotechnical plans of individual sites. The UNESCO publication illustrates fifteen examples of maps from Europe and North America (see Plate 4*a*), together with comprehensive legends and critical comments. Although this is the more recent publication it does not give the same guidance and recommendations on layout and symbols as its British equivalent. However, it does give a further classification of the lithological subdivisions or taxonomic units which can be portrayed on maps of different scales, as follows.

(1) *Lithological suites* (scales < 1:200 000) with boundaries based on the interpretation of existing geological maps or photogeology. Evaluation is based on known rock types.
(2) *Lithological complexes* (scales 1:10 000–1:200 000) with mapped boundaries which separate genetically related lithological types. Evaluation based on field and laboratory testing.
(3) *Lithological types* (scales 1:5000–1:10 000) with detailed mapping of boundaries and petrographic investigation. Rock testing.
(4) *Engineering geological types* (scales > 1:5000) Further mapping of physical state (including weathering). Evaluation of mechanical properties.

This classification basically comprises the broader small-scale regional maps (1 and 2) and the large-scale engineering plans (3 and 4) of the British Working Party report.

4.2 The publication of engineering geological maps in Britain

The British Geological Survey recognised a need for specialised maps to meet the requirements of planners and engineers some years ago, although blanket coverage of Britain was judged impracticable. One or two pioneering sheets were published in the early 1970s, starting with the 3 inches to 1 mile (1:20 000) engineering geology sheet of Belfast (Northern Ireland) in 1971, on which additional engineering orientated geological data were printed. This was largely based on records of the numerous boreholes which had been put down around the city and included an explanatory legend.

As borehole data banks were set up from the results of site investigations, and information was recorded from mines and excavations, further projects were initiated, notably at new towns such as Milton Keynes in Hertfordshire, Irvine in Ayrshire and Glenrothes in Fife (where no less than 27 maps on a scale of 1:25 000 formed an integral part of a report on environmental geology in 1982). Other similar projects have concentrated on development areas such as South Essex (1975), Cromarty Firth (1980), Southampton (1982) and Airdrie-Coatbridge (1986).

A three-phase report on the Environmental Geology of Glasgow started in the east end (Sheet NS 66) in 1983, with seventeen separate sheets including many aspects of engineering and environmental interest. Most were at a scale of 1:10 560 but some were reduced to 1:25 000. The later phases tackled West Glasgow (NS 56) and suburbs (parts of NS 55, 65 and 57) but were more selective in aspect and coverage. Short accounts[7] accompany each phase.

In IAEG terms the Glasgow maps may be described as multi-purpose, analytical and medium-scale, whereas a later survey of the Upper Forth Estuary[2] (see Plate 3*a–d*) combines the analytical approach with the comprehensive, by summarising the former in some of the published sheets. The eight maps of the upper Forth Estuary, published in full colour at a scale of 1:50 000, cover a large enough area (450 km^2) to be valuable in regional planning and yet applicable for specific site location. The report is full of useful geotechnical information and for the first time deals with the problem of depicting soil succession

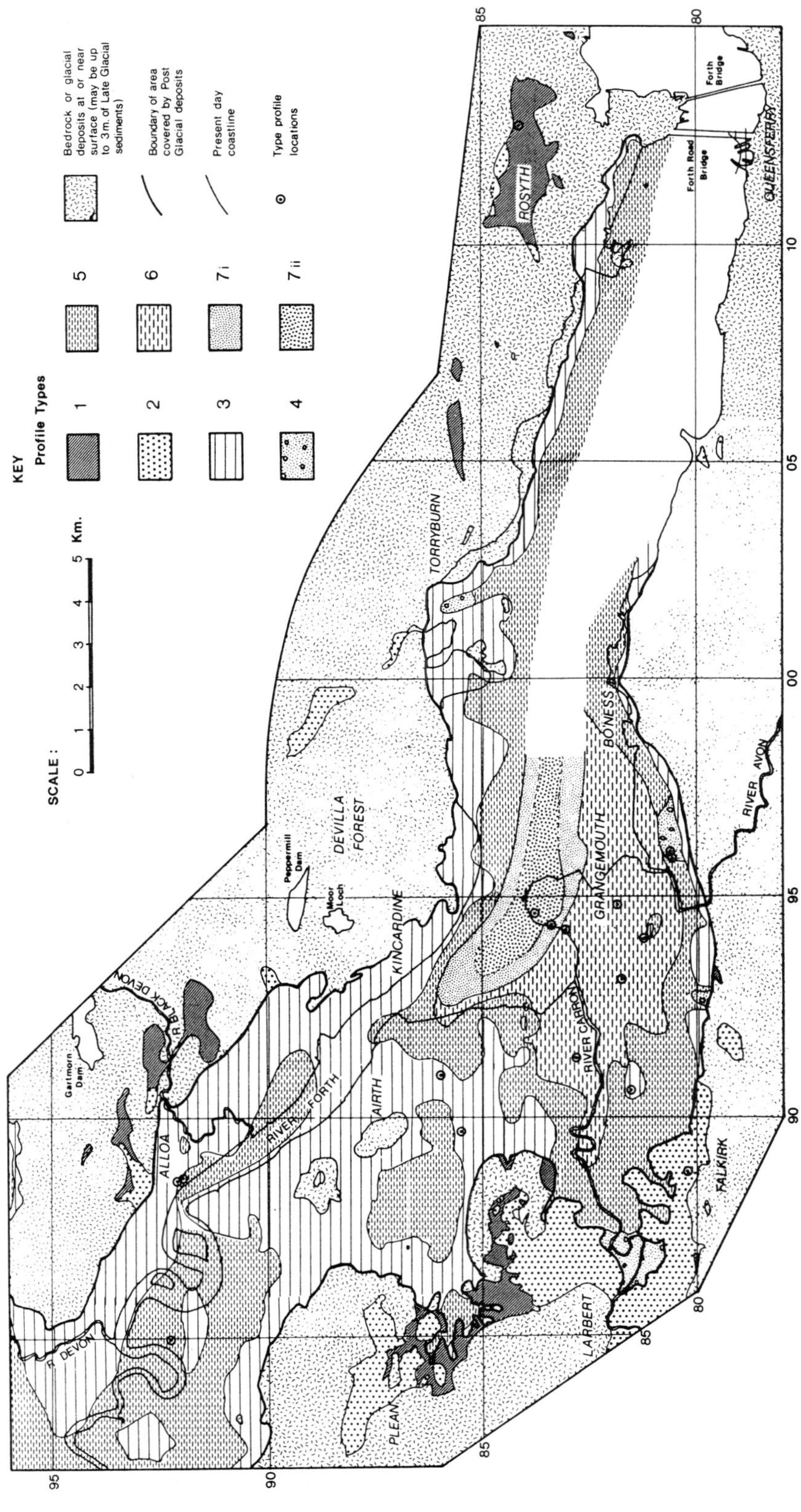

Figure 4.2 Simplified map of the upper Forth estuary, showing areas of similar drift succession. See Figure 3.9 for profile types. (From Gostelow and Browne, 1986.[2])

on maps by recognising profile types in areas where the soil succession and parameters are similar (Figures 3.9, 4.2). In order to do this, a large data base is necessary and the recent geological history of the area needs to be well understood. It is therefore doubtful whether the same types of maps could be compiled for the many parts of Britain where recent geological history is less well established.

The comprehensive nature of this study is best understood if a brief description of the available maps and sections is outlined.

4.2.1 *Maps in colour*

(1) *Engineering Geology of the Solid Rocks*: A very comprehensive clear map with useful cross-hatching to indicate mine workings, shallow rockhead, thin till and the location of the main reference points (see extract Plate 3*a*).
(2) *Drift Thickness Contours* (*depth to rockhead–isochores*): These are useful to planners and engineers and seem to have replaced the usual rockhead contours based on Ordnance Datum.
(3) *Contours to the Upper Surface of Glacial Deposits*: These are also useful since they give depth to the more consolidated material in the area (see extract Plate 3*b*).
(4) *Distribution of Mine Workings*: The workings are classified stratigraphically rather than by date or depth of extraction which could form useful additional information.
(5) *Drift Geology*: Very similar to the 1:50 000 Drift sheets of the area.
(6) *Engineering Classification of Surface Sediments*: The British Soil Classification System is used to effect throughout, although the same drift boundaries are used as on map 5 (see extract Plate 3*c*).
(7) *Geotechnical Cross-Sections*: Vertically exaggerated (× 25) sections with inferred boundaries at greater depths, below the limit of borehole control.
(8) *Geotechnical Planning Map for Heavy Structures*: This is a comprehensive map coloured in terms of a ground classification scheme graded from A to F according to site quality factors which include bearing capacity, foundation depth and the presence of shallow workings. It also gives guidance regarding the main site investigation requirements for the different grades of the classification (see extract Plate 3*d*).

4.2.2 *Monochrome maps*

Monochrome maps and sections in the text are also useful. They include:

(1) distribution of sub-surface data;
(2) simplified cross-sections of Grangemouth (see Figure 3.9);
(3) individual geotechnical profile types;
(4) map of similar profile types (Figure 4.2);
(5) thickness of Post- and Late-Glacial sediments;
(6) individual rock zone profiles.

What has been achieved by Gostelow and Browne may be repeated and enlarged upon for individual extended sites. A proposed large site which has been well investigated could have similar detailed plans included as part of the ground investigation report.

4.3 Engineering geological plans

At scales greater than about 1:10 000 many topographic features, such as roads and buildings, can be presented at their scaled size, rather than simply as symbols. Site plans are often at scales between 1:2500 and 1:500 which are demanding in view of the uncertain locational accuracy of most geological and geotechnical data. It does, however, enable macrofabrics, such as discontinuities, lithological details and/or weathering grades to be plotted much more accurately where natural exposures or open excavations are available. At greenfield or other sites lacking in rock exposures, a thick soil cover may prevent accurate prediction of rock boundaries at rockhead even where boreholes, or geophysical traverse spacings are close. In rural areas geomorphological (or landform) maps often help the soils interpretation and aerial photography may be a valuable tool. Man-made landfill areas should be investigated with the aid of older topographic maps and photographs which sometimes also provide clues to former uses of the site.

Large-scale plans compiled during, or after excavation, form a valuable record of the construction stage of a project and may enable remedial or preventative measures to be accurately specified.

4.4 Types of specialist maps and plans

Although it is beyond the scope of this book to detail all types of specialist methods for recording geotechnical data on maps, the civil engineer should have some knowledge of the value and limitations of most of these methods. It is proposed to outline the use of surface data obtained from remote sensing, and other mapping techniques, followed by a more site specific approach involving some of the techniques already alluded to earlier in this section. Finally, the mapping of environmental problems associated with landfill or subsidence hazards and groundwater pollution are considered.

4.4.1 *Surface evaluation*

Whilst *topographic* maps and plans should not require further explanation, the value of consulting old maps cannot be overstated, particularly in connection with sites formerly used by man. Old shafts, quarries and changes in natural features may all become apparent if early maps, air photographs or even surface photographs can be discovered. When used in conjunction with the geological maps already described, there may be sufficient information available to initiate a more effective site investigation. In certain circumstances, initial research may involve the use of one or more of the many methods of *aerial imagery* (remote sensing) now available worldwide.

In developed countries, large-scale aerial photography may be most appropriate and available, whereas in remote or large countries, perhaps lacking in comprehensive map coverage, *satellite imagery* may be more applicable for the purposes of terrain evaluation in the first instance. It follows that the use of remote sensing by engineers can be summarised under two headings: (1) the use of aerial photography as an aid to producing larger scale topographical, geomorphological and geological maps and plans, and (2) the use of satellite imagery in reconnaissance and initial evaluations over large areas. Since much of the multi-spectral analysis is carried out on satellite images, these and other techniques will be considered in a later section.

Aerial photography. The most common type of remote sensing used in civil engineering is based on high quality black and white or colour (occasionally false colour, enhanced towards the near infrared) vertical photographs taken from an aircraft. The most common photographs are printed from large format (230 × 230 mm) negatives or 70 mm film. Since the normal focal length of lenses for this type of work is 152 mm (6 in), the scale of the final photograph is approximately twice the height (in feet) of the sortie run (e.g. a 1:3000 photograph requires an average flying height above the ground of about 1500 ft). Whilst more stable twin-engined aircraft are most commonly used for photogrammetry, experiments have proved reasonably successful using portable radio-controlled model aircraft or easily transported microlite aircraft, to obtain photographs for interpretation purposes only. For photogrammetric purposes it is necessary to fly runs at a constant height and heading to produce photographic images which have a 60% overlap. This gives the added advantage of stereoscopic (3D) viewing. If additional survey control markers are set on the ground before the sortie, photogrammetric plotting can be used to produce accurate site plans with contours at 0.5–1 m intervals, depending on the scale. There are specialist firms which can provide excellent service with high quality cartography and photopositives (see Appendix D for British addresses).

In certain areas of high relief it may be useful to carry out a run using oblique photography so that a steep hillside or cliff (which may not show up well on plans or vertical photos) can be

studied stereoscopically. Terrestrial photogrammetry of inaccessible rock faces has also proved invaluable, for example in the design of stabilisation work.

One of the main advantages of photogeological interpretations is that *ground texture* can be plotted. This is an important feature unobtainable from most published maps for it shows exposures, rock structure, land forms of all types, drainage details and vegetation types. It also enables the influence of man on the landscape to be readily assessed (see Figure 4.3).

With initial identification of these features based on stereoscopic viewing, it is then possible using the less distorted central areas of photographs to compile a first draft geological/geomorphological map which should prove a valuable base for a walk-over survey, or a more accurate and detailed geomorphological mapping programme.

Photogeological work for proposed road schemes has been recommended in Britain by, amongst others, the Transport and Road Research Laboratory since the early 1970s. It has been used throughout the world to good effect and is especially useful where low angle sun highlights undulations and other significant features such as hollows and slight scarps in relatively flat land. In hill areas, too much shadow can obscure detail and photographs taken nearer to the middle of the day are recommended.

It is not only road schemes which can benefit from aerial photography. The proposed sites of dams and reservoirs, large building complexes, new towns and coastal works can all be studied to advantage in this way.

If a series of photographs taken on well separated dates is available, the effects of landfill, subsidence, landslip or erosion may be detected. One of the most convincing uses of periodic air photography of this type was made in connection with the 1969 analysis of the Aberfan coal tip disaster in South Wales. Both the build up of the tip and its effect on local drainage could be seen on several photographs taken between 1945 and 1965. Perhaps local authorities should be commissioning large-scale periodic photography of known suspect ground as a matter of routine surveillance.

As far as detailed geomorphological work is concerned, it is possible to detect minute features from good quality photographs only if there is sufficient lack of tall and dense vegetation. Engineers should therefore try to plan coverage of proposed sites during winter months if deciduous woodlands are involved.

Without too much vegetation cover, former positions of river channels in flood plains become clearly visible. This is even more true of *periglacial* (permafrost) and *solution* features which are often clearer on a photograph than to an observer standing on the ground! To an engineer looking for borrow material, the extent, volume and shape of glacial and fluvio-glacial landforms, alluvial fans and other gravel banks are important features. Accurate plotting of landslips and changes of slope can be combined with field measurements of slope angles using an Abney level or sighting clinometer. Other observations which may have remained undetected from the photography can also be added to form an engineering *geomorphological map*. An example of a geomorphological map for engineering use is given in Figure 4.4. In some circumstances geomorphological maps showing fairly static conditions may be supplemented or even replaced by *geodynamical plans* which indicate more active or ephemeral features.

Black and white infrared and false-colour infrared photography have their places in photogeological interpretation since they greatly enhance contrast and increase penetration of any haze which may be present. Water and moisture in soils are more absorbent to near infrared and appear black on prints. Various types of vegetation show up in tones of red due to variations in reflectance between species. If both near infra-

Figure 4.3 Aerial stereo pair at a scale of 1:10 000 from Crummackdale, North Yorkshire (Meridian Airmaps, Photographs 4868 074 and 4868 075). View with a 70 mm hand stereoscope for a three-dimensional image of ground texture, outcrops, structures and slopes. Use acetate overlay to record details.

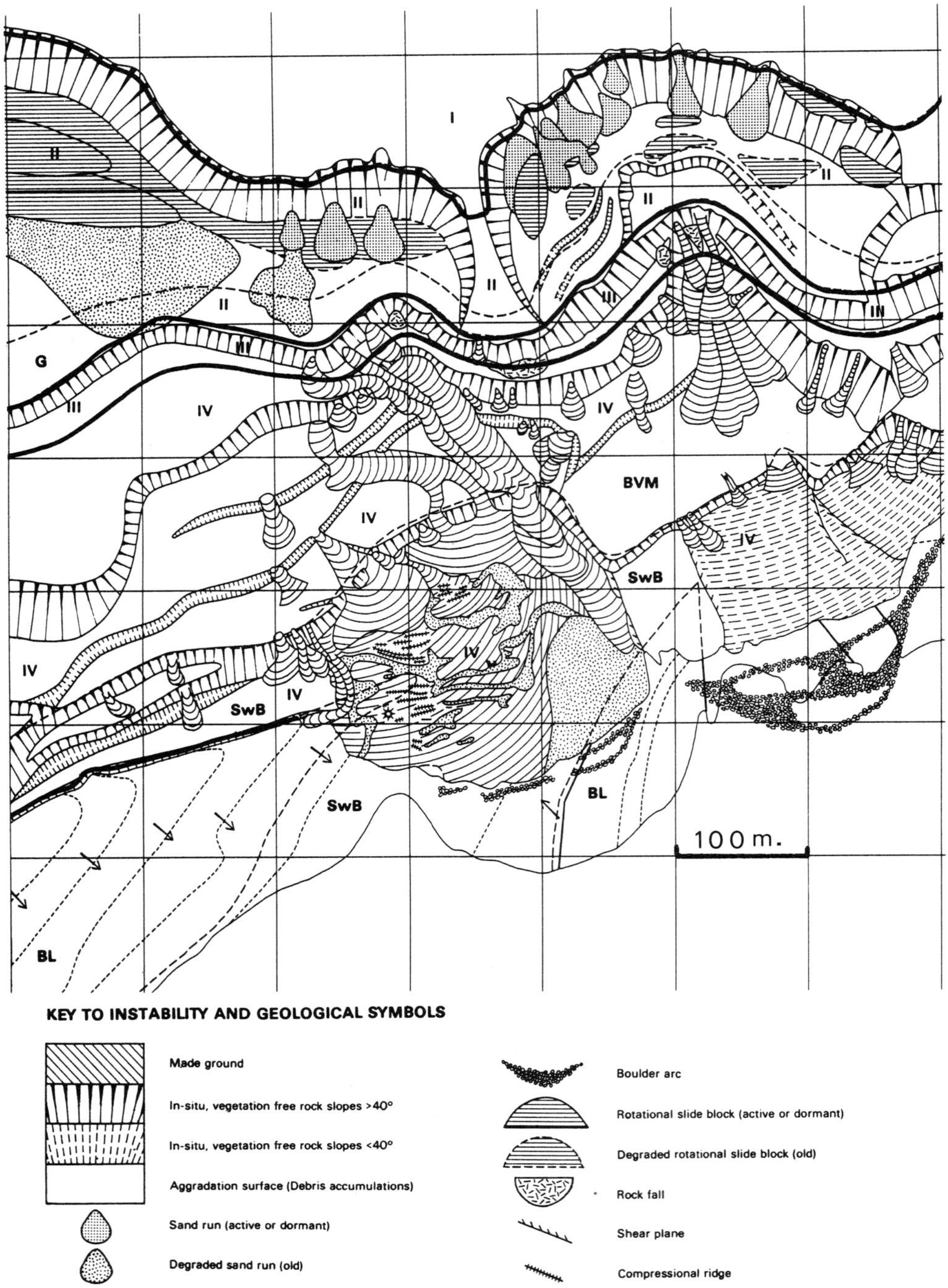

Figure 4.4 Geomorphological and terrain evaluation map. Part of a published BGS map of the Dorset coast. (Reproduced by permission of the Director, British Geological Survey. Crown/NERC copyright reserved.)

red and normal colour or black and white coverage is available for a site then maximum advantage will be obtained.

Satellite imagery. Satellite coverage is now becoming more widespread, with definition improving all the time. Most of the commercial images from space have been from the US Landsat systems with pixel definition up to 30 m square on the Thematic Mapper. This type of television image has recently been improved in the French SPOT satellite, which is providing a pixel size nearer to 10 m square. However, even with this definition there is some limitation on the satellite image for civil engineering use. It can usually be bettered by aerial photography and Landsat lacks the very important stereoscopic facility. On the other hand, its value for the mapping and physical evaluation of remote and often unmapped countries is tremendous. Even with a resolution of 30 m on the latest Landsat 4 Thematic Mapper, the small-scale infrared photo images are impressive.

The main advantage over normal aerial photography is the facility of being able to use up to seven bands of infrared waves which can be 'mixed and matched' to pick out various aspects of the earth's surface. The images obtained can be computer processed to remove unwanted contrast and striping or to add classification to images using density slicing and digital filtering. This is a type of radiance contouring system. Band 7, with an electromagnetic wavelength of 2.08–2.35 μm has proved to be the most discriminating of the infrared bands as far as lithological and hydrothermal features are concerned. However, materials can sometimes be successfully recognised using thermal infrared line scanning techniques (8–14 μm) which depend on radiant temperature and emissivity of rocks and soils. The thermal inertia depends on conductivity and density so that diurnal differences give a clue to surface materials. Shales, gravels and sandy soils display rapid diurnal changes, whereas quartzites and dolomites stay relatively unchanged from day to night. Moisture conditions in the ground show up and, in certain circumstances, joints and sinkholes in dolomite invisible on normal panchromatic photographs, can be distinguished. The method has also been attempted in temperate climates with some success.

One of the major problems of satellite imagery is cloud cover, which may be persistent in equatorial countries. To overcome this problem *radar systems* have been developed which can be used in all weather, day or night situations. Using a side-scan technique it is possible to obtain images which can be used in landform recognition, although this does not yet appear to be commercially viable in civil engineering.

Terrain evaluation. With the aid of either high level aerial photography or enhanced satellite images on scales between 1:250 000 and 1:100 000 it is possible to classify landscapes in various ways. Landscape-based terrain evaluation techniques may divide an area into *facets* each consisting of a number of similar geomorphological or geological units which characterise the facet. This method has been successfully applied in South Africa where terrain evaluation forms an essential aspect of site investigation planning.[8]

An alternative *parametric* classification concentrates more on individual geotechnical parameters which may be essential to a particular project. For example, in developing countries where climatic conditions and simpler design specifications may control new roadworks, efficient use of local materials may be facilitated by early application of indirect reconnaissance based on remote sensing. In this type of work particular parameters such as slope, type of material and quantities of cut and fill may be estimated at the outset. Interpretation would then be followed up by ground checking and, if necessary, some site investigation.

4.4.2 *Geotechnical appraisal of rock masses*

Once selected sites are excavated for foundations, cuttings, slopes or tunnels, the detailed rock mass characteristics become extremely important, so that the results of any further investigations need to be plotted on large-scale plans and sections. Conventional site plans of geological conditions derived from boreholes, trial pits and natural exposures usually concentrate on the main distribution of *lithological types*. Whilst this sort of plan is useful it may not include the most vital aspects of the rock mass which affect ease of excavation or even stability. Geotechnical parameters such as strength, discontinuity orientations and types, weathering grades or alteration states may not all be recorded sufficiently at the site investigation stage despite the fact that they form a vital aspect of the design of the works.

If site investigation has been carefully planned and is comprehensive, it should be possible to highlight particularly important parameters by using specialised geotechnical plans and sections. Examples of some types of geotechnical plans have already been given in previous sections as follows:

(1) rockhead contour plans;
(2) contours on particular horizons (structure contour plans);
(3) contours at equal depths from the surface (isobath plans);
(4) contours showing thickness variations (isopach plans).

Most of these plots require numerous boreholes, but they do help to define problems such as buried channels, old infilled quarries, rockhead highs, thickening soil units or perhaps depth and geometry of worked coal seams.

The combined use of surface exposures and borehole data should be common practice, although the two are often given separate consideration by engineers and surprisingly the former is sometimes regarded as being outside the scope of a site investigation contract. Rather than plot surface data at the pre-tender stage there has, in the past, been more emphasis placed on the results of boring and drilling. If large-scale plans of sufficient accuracy are available on which to plot local data, much more may be obtained from a rock mass, soil bank or trench exposure than from a small diameter borehole core or sample. However, in poorly exposed ground a downhole CCTV method may be employed to assess the in-situ discontinuity situation provided boreholes are taken at various orientations. Where plans are lacking, the use of photographs (with traced overlays or annotation) can give much more detail of the three-dimensional surface shapes and variations than can be obtained from boreholes alone.

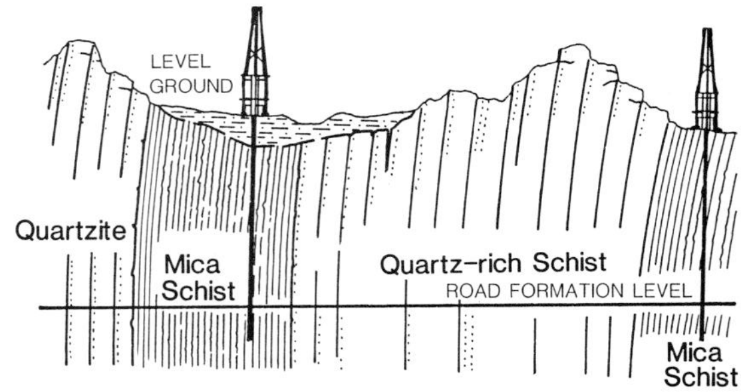

Figure 4.5 Section with vertical boreholes and near-vertical strata. A poorly planned situation where boreholes were set up on level ground (over mica schist units) and natural exposures ignored.

An example is given in Figure 4.5 where vertical boreholes, taken in a vertically dipping rock succession, were far more misleading to the contractor than a cheaper and more informative exposure plot would have been. In this example boreholes were placed at logistically convenient positions and no account was taken of the orientation of structures within exposures nearby, or even the exposures themselves. Inclined boreholes, although more expensive, would have helped to plot a more accurate section here.

At dam sites and especially where tunnels are to be constructed under mountainous terrain, the distribution of *surface exposures* and their structures is of vital importance in keeping down drilling costs and improving the quality of predictive plans and sections. In Figure 4.6, it is clear that good exposure enables some predictions to be made, but two important points arise from the section. Firstly, a borehole is essential to find the level of the unconformity in the tunnel and the ground water conditions above it, and secondly, where superficial deposits obscure the ground, the worst conditions should

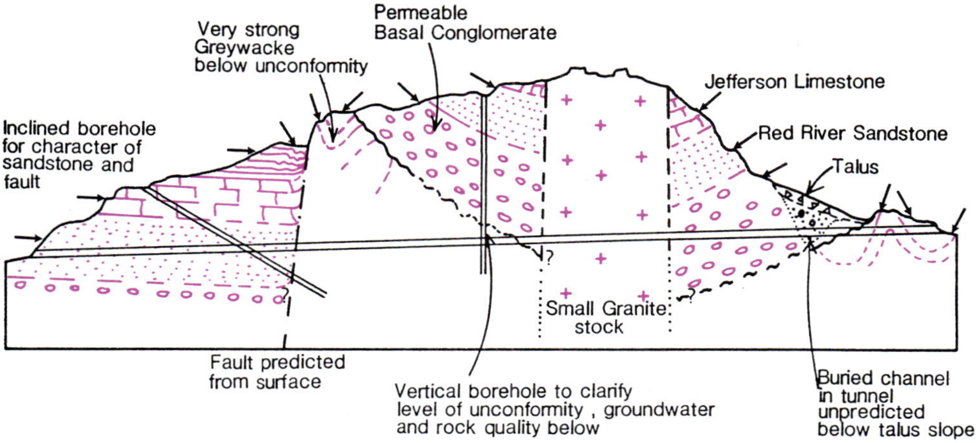

Figure 4.6 Surface mapping used to predict invert conditions in a tunnel section. Boreholes from the mountain top are justifiable if major uncertainties arise in the prediction.

be expected and may have to be proved not to occur.

Rock mass data at the site investigation stage. Natural rock exposures vary in quality depending on size, state of weathering and vegetation cover. Sometimes only limited data can be retrieved from a small, lichen covered and strongly weathered corestone whereas a freshly eroded or blasted, clean expanse of rock can sometimes give an overwhelming amount of detail.

At each exposure, the numerous observations may be logged rapidly using a clinometer and compass or a 'Clar' type compass especially designed for dip and strike measurements on discontinuities. They can then be recorded together with a field sketch and plotted directly on a plan. Alternatively they may be combined with a rock mass description data sheet and discontinuity survey data sheet (Figures 4.7, 4.8), as recommended by the Engineering Group Working Party report on 'the description of rock masses for engineering purposes'.[9] The data sheets are used to provide rock mass characteristics in clearly defined standard terms which civil engineers can refer to with some confidence. The results of further selected field tests, such as the 'L' type Schmidt hammer, may be included in the assessment and weightings assigned to the more important and persistent fractures. A method used by geologists to solve geometrical, statistical and stability problems is the stereographic projection (see Appendix C), computer packages for which are also available.

Rock mass data at the construction stage. Various methods of plotting rock mass data can be employed for different purposes at this stage. For example, when tendering, during construction (for recording what was found) and following excavation (for specifying remedial measures).

When *tendering*, the estimator needs as much information (preferably in a simplified but representative form) as he can obtain within the time available. This is most commonly provided in the form of a site investigation report but may be supplemented by access to cores, plans and other reports. A good engineer will have foreseen most of the problems associated with the earthworks and will provide the necessary information to help the contractor execute the works.

A simple form of rock and soil classification (Quality or Grade) may be possible from accurately logged and closely spaced boreholes. Various methods have been applied in the last 20

ROCK MASS DESCRIPTION DATA SHEET

GENERAL INFORMATION

Seq. No. | Site | Date (Day, Month, Year) | Operator | Method of location | Co-ordinates or chainage (metres): Northings or chainage, Eastings, Elevation

Method of location:
1. By co-ordinates.
2. Chainage
3. On attached map/ drawing/photograph

Locality type
1. Natural exposure
2. Construction excavation
3. Trial pit
4. Trench
5. Adit
6. Tunnel

Size of locality
1. > 10m²
2. 5-10m²
3. 1- 5m²
4. < 1m²
5. Line survey

No. of supplementary sheets of discontinuity data

Sketch | Photograph
0. No
1. Yes

Field tests
Specify type

Remarks

ROCK MATERIAL INFORMATION

Colour
1. Light
2. Dark

1. pinkish
2. reddish
3. yellowish
4. brownish
5. olive
6. greenish
7. bluish
8. greyish

1. pink
2. red
3. yellow
4. brown
5. olive
6. green
7. blue
8. white
9. grey
0. black

Grain size
1. Very coarse (>60mm)
2. Coarse (2-60mm)
3. Medium (60μ-2mm)
4. Fine (2-60μ)
5. Very fine (<2μ)

Compressive strength
1. Very strong (>100MN/m²)
2. Strong (50-100MN/m²)
3. Mod. strong (12.5-50MN/m²)
4. Mod. weak (5-12.5MN/m²)
5. Weak (1.25-5MN/m²)
6. V. weak/hard (600-1250kN/m²)
7. Very stiff (300-600kN/m²)
8. Stiff (150-300kN/m²)
9. Firm (80-150kN/m²)
0. Soft (40-80kN/m²)

Method of determining compressive strength
1. Measured
2. Assessed

Rock type

Qualifying terms to describe rock

ROCK MASS INFORMATION

Fabric
1. Blocky
2. Tabular
3. Columnar

Block size
1. Very large (>8m³)
2. Large (0.2-8m³)
3. Medium (0.008-0.2m³)
4. Small (0.0002-0.008m³)
5. Very small (<0.0002m³)

State of weathering
1. Fresh
2. Slightly
3. Moderately
4. Highly
5. Completely
6. Residual soil

No. of major discontinuity sets

LINE SURVEYS TO DETERMINE DISCONTINUITY SPACINGS

	Plunge of line	Trend of line	Length of line (metres)	No. of fractures	Spacing	Remarks
Line 1						
Line 2						
Line 3						

Discontinuity spacing
1. Ext. wide (<2m)
2. Very wide (600mm-2m)
3. Wide (200-600mm)
4. Mod. wide (60-200mm)
5. Mod. narrow (20-60mm)
6. Narrow (6-20mm)
7. Very narrow (6mm)

Figure 4.7 Rock Mass Description Data Sheet. (From Working Party Report.[9])

DISCONTINUITY SURVEY DATA SHEET

GENERAL INFORMATION

Seq. No. | Site | Date (Day, Month, Year) | Operator | Discontinuity data sheet No. of

NATURE AND ORIENTATION OF DISCONTINUITY

Chainage or No.	Type	Dip	Dip direction	Persistence	Aperture	Nature of infilling	Consistency of infilling	Surface roughness	Trend of lineation	Waviness wavelength	Waviness amplitude	Water/Flow	Remarks

Type
0. Fault zone
1. Fault
2. Joint
3. Cleavage
4. Schistosity
5. Shear
6. Fissure
7. Tension crack
8. Foliation
9. Bedding

Dip, Dip direction and Trend of lineation
(Expressed in degrees)

Persistence
(Expressed in metres).

Aperture
1. Wide (>200mm)
2. Mod. wide (60-200mm)
3. Mod. narrow (20-60mm)
4. Narrow (6-20mm)
5. Very narrow (2-6mm)
6. Ext. narrow (<2mm)
7. Tight

Nature of infilling
1. Clean
2. Surface staining
3. Non-cohesive
4. Inactive clay or clay matrix
5. Swelling clay or clay matrix
6. Cemented
7. Chlorite, talc or gypsum
8. Others - specify

Compressive strength of infilling
1. Very soft (<40kN/m^2)
2. Soft (40-80kN/m^2)
3. Firm (80-150kN/m^2)
4. Stiff (150-300kN/m^2)
5. Very stiff (300-600kN/m^2)
6. Hard/v.weak (600-1250kN/m^2)
7. Weak (1.25-5MN/m^2)
8. Mod. weak (5-12.5MN/m^2)
9. Mod. strong (12.5-50MN/m^2)
10. Strong (50-100MN/m^2)
11. Very strong (100-200MN/m^2)
12. Ext. strong (>200MN/m^2)

Roughness
1. Polished
2. Slickensided
3. Smooth
4. Rough
5. Defined ridges
6. Small steps
7. Very rough

Waviness
Express wavelength & amplitude in metres

Water
1. Dry
2. Seepage

Flow
3. <10ml/sec
4. 10-100ml/sec
5. 0.1-1 l/sec
6. 1-10 l/sec
7. 10-100 l/sec
8. >100 l/sec

Figure 4.8 Discontinuity Survey Data Sheet. (From Working Party Report.[9])

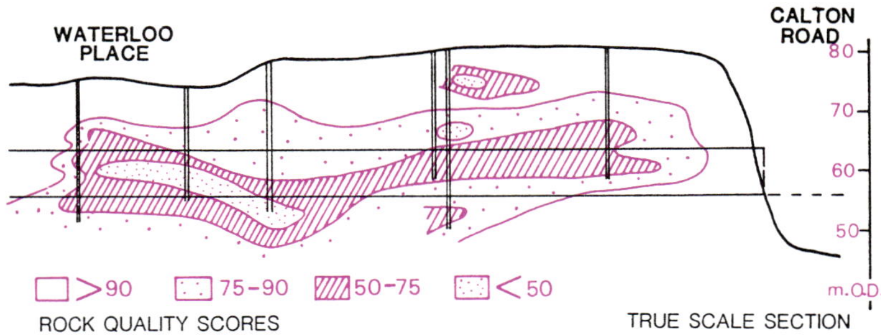

Figure 4.9 Relative hardness based on rock quality scores from boreholes for the Calton Hill tunnel, Edinburgh. (After Cottiss *et al.*, 1971.[11])

years or so (see Figure 4.9), but in practice borehole data are often limited and only simple rock gradings can be achieved.

The main quantifiable characteristics of rock commonly shown on engineering borehole records are (1) unconfined compressive strength, (2) weathering grade and (3) rock quality designation (RQD) or preferably lithology quality designation (LQD) based on joint spacings in different lithologies.[10] The author has found it useful to combine these limited data as follows:

Value of compressive strength, $(C_u) = 0\text{–}200$ MPa
Value of LQD% × 2 $= 0\text{–}200$
Increments of weathering grades $(W) = 0\text{–}100$
$(WI = 100,\ WII = 80,\ WIII = 60,\ WIV = 40,\ WV = 20,\ WVI = 0)$

If the rock grade number,

$$\text{RGN} = C_u + 2\text{LQD} + W,$$

then the higher the score up to a maximum of 500, the more difficult it is to excavate the rock. This may present the estimator with a useful guide in tendering for small tunnel and highway contracts, but it is *not* intended that this should form a basis for any later claims. The range of scores may be classified A–D on the following basis:

Grade A	> 400	Blasting may be required
Grade B	200–400	Difficult without blasting
Grade C	100–200	Blasting not normally required
Grade D	0–100	Easily removed

These classes of rock, which are not unlike those used by Cottis et al.[11] (Figure 4.9), may then be plotted on a section. For tunnels, the *scaled ground log* is another way the author has found to use similarly limited site investigation data in a form which could be of value when estimating risk (see Figure 4.10).

During construction, plans are updated and records are kept showing where rock masses have been exposed by excavation and their true state of weathering etc. has been determined at known levels.

In open excavations during construction it is not always the lithological distributions which are the controlling factors. In some tropical and sub-tropical zones, the chemical weathering process can be represented by variations in the rock mass weathering grades I–VI as defined in BS 5930:1981 (Site Investigations).[12] However, most surveys at the construction stage normally involve a combination of rock mass characteristics so that Rock Qualities or Grades can be defined. The most commonly quoted pioneering work on both weathering grades and the rock gradings concept is that of Knill and Jones[13] which includes weathering zones in gneisses at the Roseires dam on the Nile and rock grades at the Latiyan dam in Iran (Figure 4.11).

At the Latiyan dam, rock grades I–VII varied

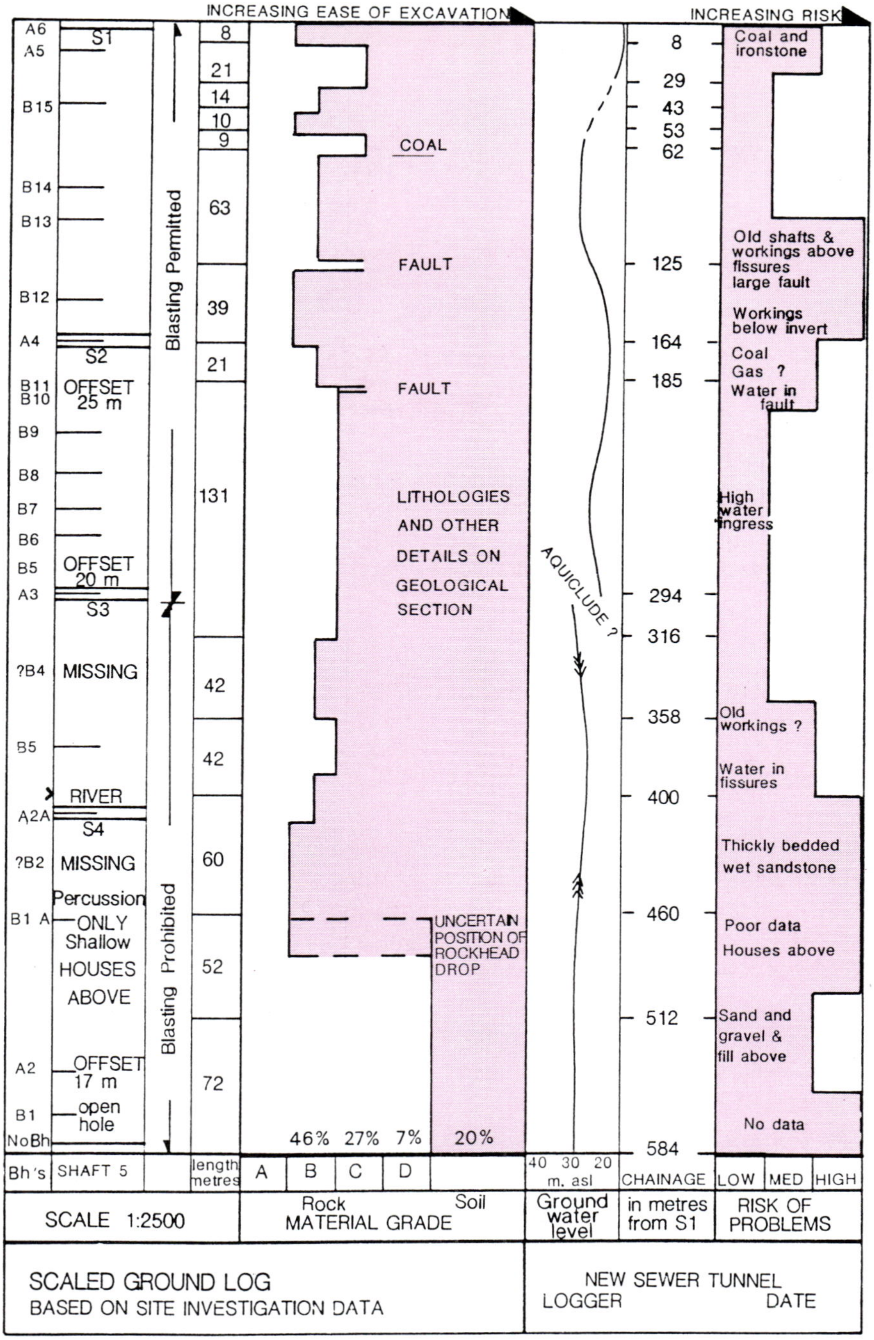

Figure 4.10 Summary of tunnel conditions prepared for an estimator at the tendering stage of a tunnel contract.

from sound massive rock with no weak seams and widely spaced joints (Grade I) to friable clay shales (Grade VII). The grading classification was based on:

(1) state of weathering and loss of cohesion;
(2) relative compactness of the rock;
(3) intensity and orientation of various sets of fractures;

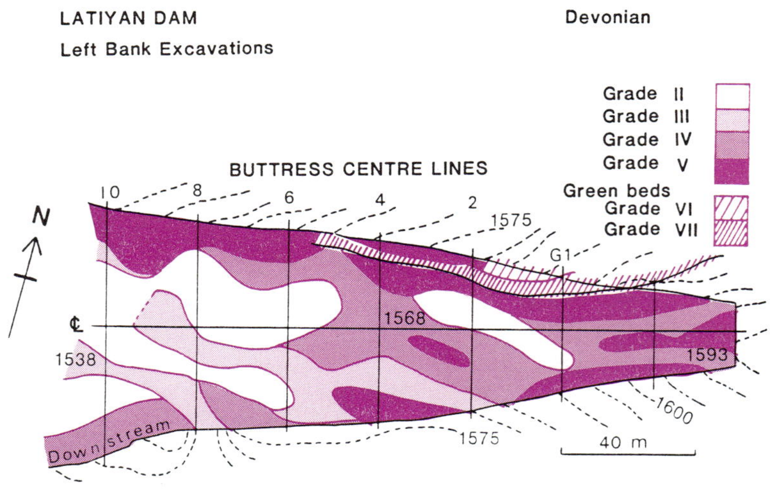

Figure 4.11 Variations in rock conditions in the excavated foundations of the Latiyan dam, Iran. This geotechnical plan illustrates the use of a rock grading scheme. (After Knill and Jones, 1965.[13])

(4) relative cleanliness of fractures;
(5) relative abundance of shale layers.

These factors were site-specific and obtained from open excavations. They were correlated with the results of geophysical work and tests in boreholes, which helped to assess the engineering behaviour of the dam foundation.

Subsequently rock structure classifications have been developed, mainly for establishing ground support requirements in tunnelling.[14] Bieniawski[15,16] has produced a complex system of rock classes each of which contain individual rock mass ratings (RMR). Barton[17] defined Rock Mass Quality (Q) in terms of six main parameters:

(1) RQD;
(2) J_n, the number of joint sets;
(3) J_r, the roughness of the most unfavourable set of joints;
(4) J_a, the degree of alteration of the most unfavourable set of joints;
(5) J_w, the joint water reduction factor (seepage);
(6) SRF, the stress reduction factor which account for squeezing or swelling properties.

Thus,

$$Q = \underbrace{\frac{\mathrm{RQD}}{J_n}}_{\text{relative block size}} \times \underbrace{\frac{J_r}{J_a}}_{\text{interblock shear strength}} \times \underbrace{\frac{J_w}{\mathrm{SRF}}}_{\text{active stress}}$$

Q ranges from 0.001, for exceptionally poor quality swelling rock, to 1000 for exceptionally good quality rock. Although these parameters are difficult to assess from rock cores, they may be more effectively applied to tunnels being mapped in preparation for tunnel support systems.

Recording the geology of tunnels during construction can, however, be a difficult process if linings are being placed immediately behind a full-face tunnelling machine, but if blasting is used and the lining process is some distance behind the advance of the face, recording may be accurately carried out by the use of both face and axonometric longitudinal plots where the centre line at the crown forms the centre of a strip showing both left and right walls at the top and bottom of the strip, respectively (see Figure 4.12).

Where only occasional sampling is possible, exact sample sites should be recorded and observations of the structure across the tunnel

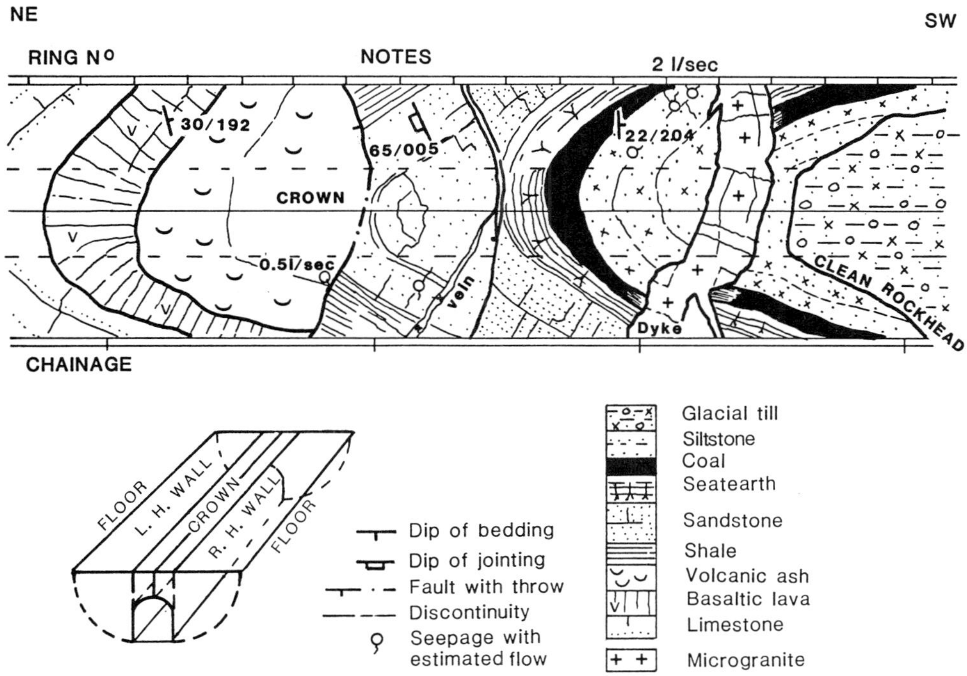

Figure 4.12 Basic layout of an axonometric plot of tunnel geology. This gives a better view of the three-dimensional structure than a simple section, and can be modified to take notes or record various parameters on strips alongside.

noted, wherever scan lines can be obtained. Figure 4.13 shows a tunnel section recorded by an engineer with little geological knowledge and the subsequent attempt by a geologist to correlate this with the site investigation results (not an easy task!).

Good accurate records help to pinpoint areas which require remedial measures such as rock bolting and grouting. This is equally true for any major rock excavation (see Figure 4.14), and may be helped by the use of carefully controlled colour photography using long focus lenses to reduce distortion. Accurate tracings, preferably on clear acetate overlays, can then be plotted to

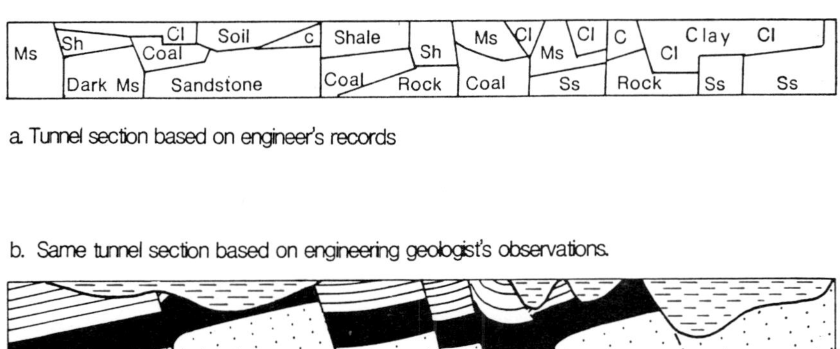

Figure 4.13 Tunnel records.

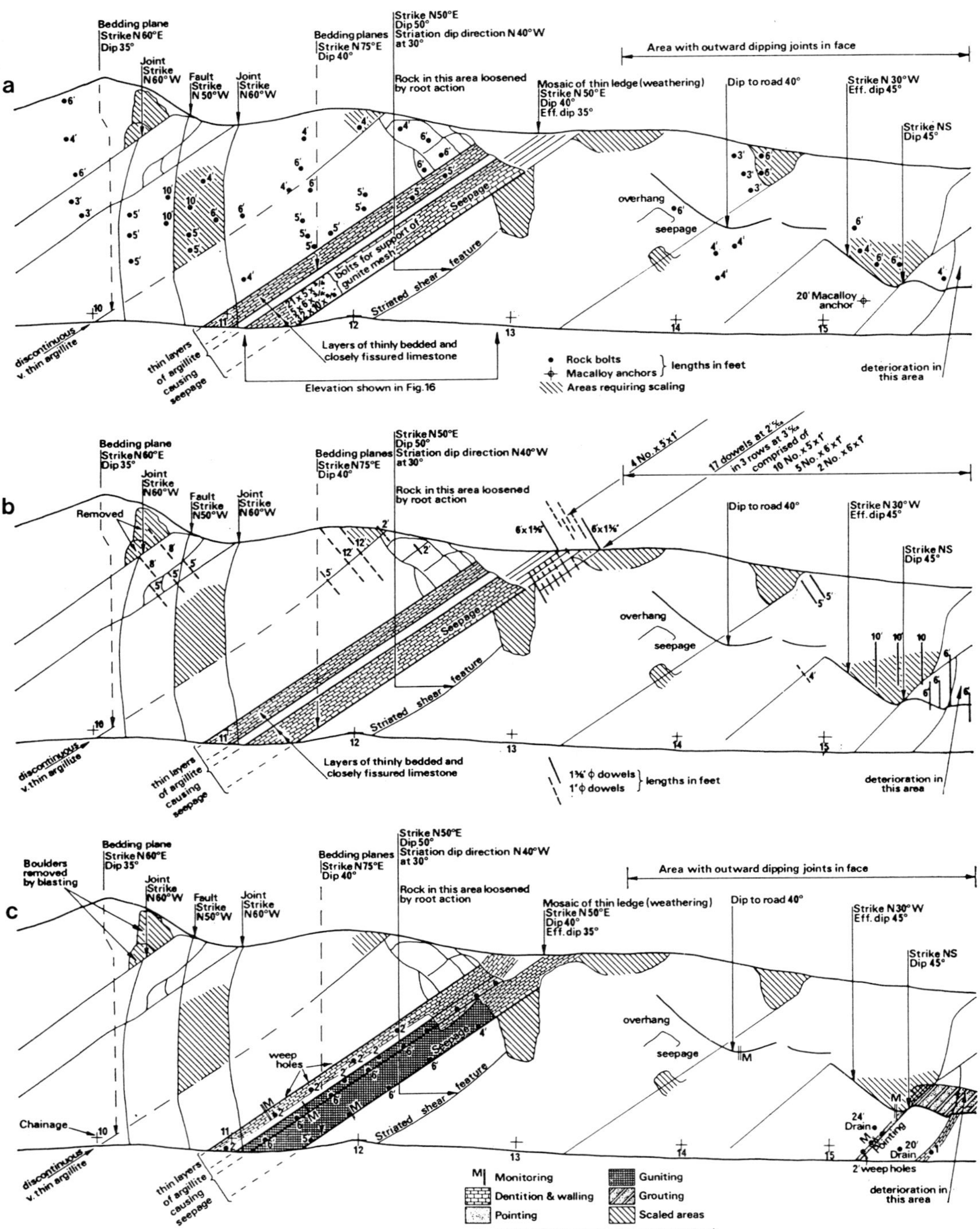

Figure 4.14 Geotechnical sections at the construction stage of the Taff Vale Trunk Road, S. Wales. (*a*) Elevation of part of the road cut face showing geological conditions, the location and lengths in feet of rock bolts and Macalloy anchors, and areas requiring scaling. (*b*) Location of dowels. (*c*) Location of monitors, and areas requiring dentition and walling, pointing, guniting, grouting and scaling. (From Dearman and Fookes, 1974.[5])

pick out lithologies and discontinuity patterns and thus help specify the necessary remedial action.

4.4.3 *Geotechnical plans for soils*

Although rocks should be mapped in detail for work involving excavation, soil distribution across a site can be even more critical. There are many more suspect soil materials which affect design procedures than is the case with rocks, though swelling, very weak, sheared or highly weathered rocks can cause major problems.

Most of the quantitative aspects of soil

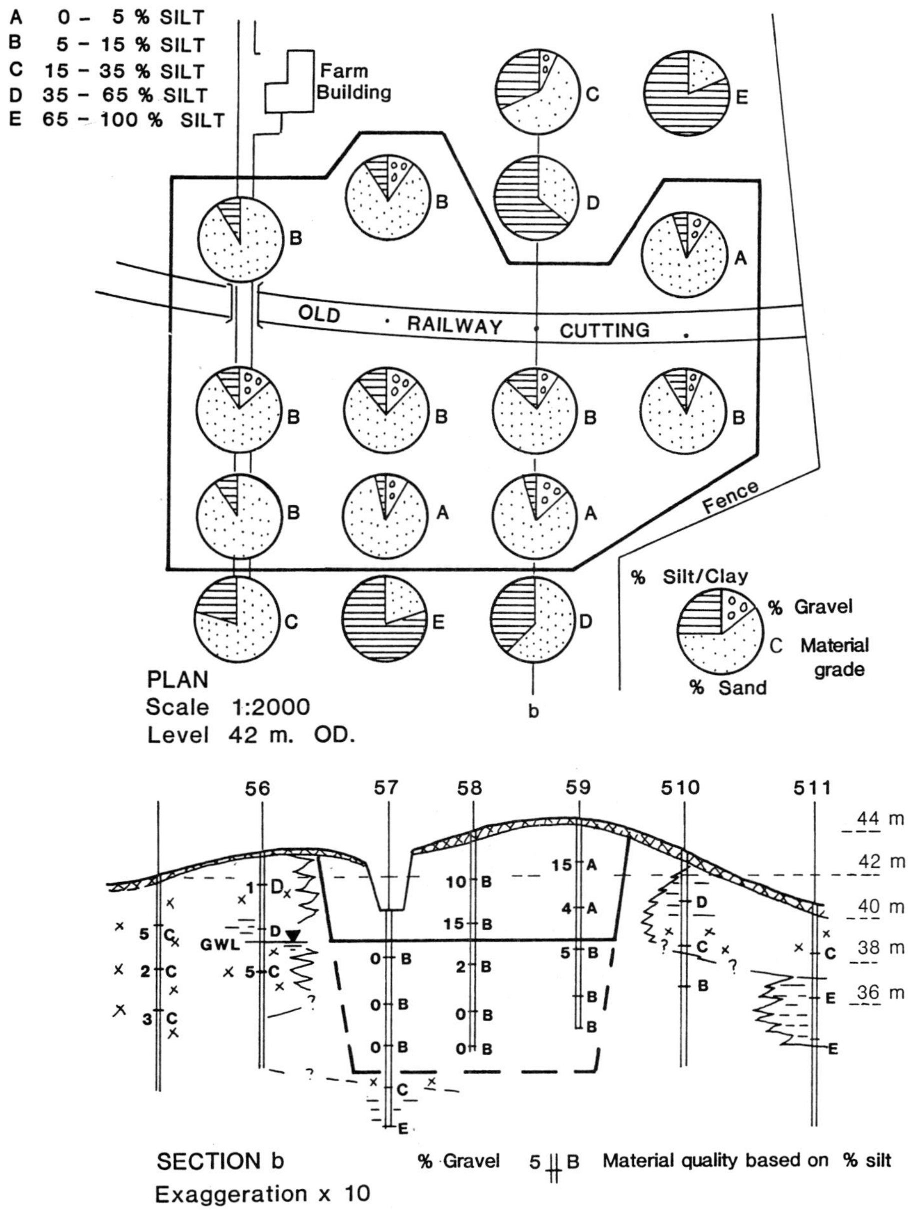

Figure 4.15 Geotechnical plan and section based on sieve analyses of granular material sampled at specific levels and quality graded A–E.

mechanics depend on strength, compressibility, soil-water relationships and other critical factors which can vary widely with both depth and horizontal distribution. Many soil mechanics calculations are based on the assumption that soil materials are semi-infinite and homogeneous so that even thin bands of contrasting materials can upset predictions. It is important, therefore, that the detailed distribution of soil parameters is obtained for design purposes, especially where foundations are based within soils or where soils are to be excavated. How can soil distribution and anisotropy be depicted for a site? Geomorphological studies can help in favourable circumstances, but in flat areas or where disturbed ground is involved, trial pit or borehole logs have to be relied upon.

Analytical plans showing single or multiple soil parameters may then be plotted to indicate degree of homogeneity across the site. The difficulties of correlation in soils have been highlighted in Chapter 3, but where borehole spacing is adequate, distribution of measured parameters obtained from samples should aid assessment. For example, Figure 4.15 shows one of a set of serial sections and plans based on soil gradings for an area of possible material extraction. However, other parameters such as strength, plasticity, permeability, or friction ratio from penetration testing, could be similarly plotted and contoured. Within superficial deposits, the best form of plan might show soil parameters taken from samples obtained at specific levels, such as formation level.

4.4.4 *Geophysical maps and plans*

It is not possible here to attempt to portray the great variety of geophysical parameters which can be plotted on maps and sections. However, some comment can be included on the types of plots most commonly encountered by engineers.

There are basically two types of geophysical measurements: those which record variations in natural fields due to differences in ground conditions and those which result from an artificial impulse created by the method being employed.

Natural fields include

(1) magnetic fields,
(2) gravitational fields.

Artificial impulses include

(1) seismic waves,
(2) electrical and electromagnetic impulses,
(3) microwaves.

Magnetic anomalies arise from the variation in magnetic susceptibilities of different materials in the ground and can be picked up using a proton magnetometer or perhaps a simpler torsion magnetometer where signals are strong. Magnetic maps contour the values of field strength (measured in gammas or nanoteslas) or magnetic gradient, and can be gathered by hand-held instruments on site to find, for example, basic igneous rocks and mine shafts (see Figure 4.16). On small-scale aeromagnetic maps, the contouring is derived from flight path recordings using a proton magnetometer and is mainly for regional use. Published aeromagnetic maps are available in Britain, as indicated in Figure 1.3. Care has to be exercised in their interpretation due to the fact that anomalies are not usually

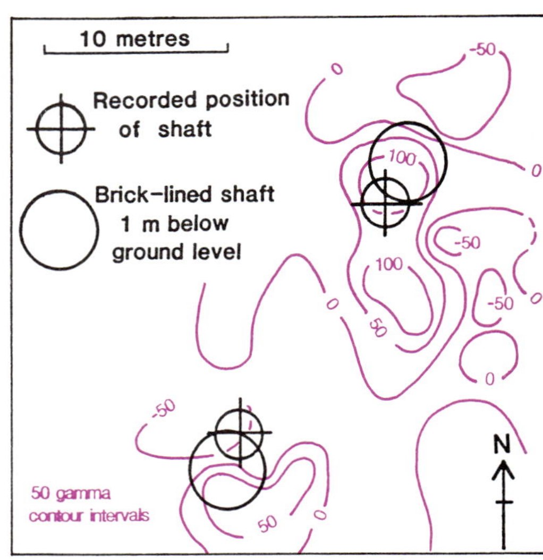

Figure 4.16 Part of a proton magnetometer survey showing the offset relationship between positive magnetic anomalies and the shaft positions as proved by subsequent excavation. (After Higginbottom.[18])

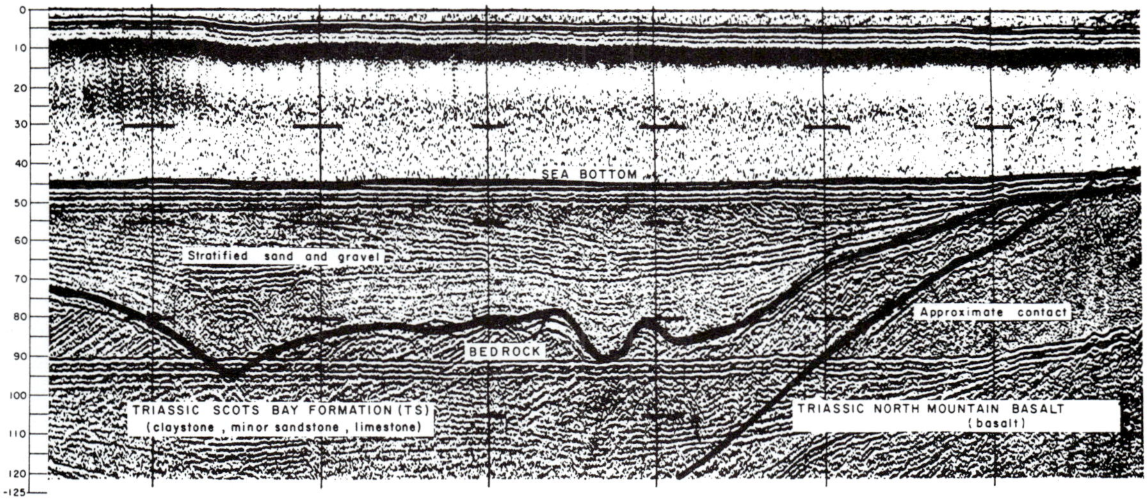

Figure 4.17 Continuous seismic reflection profile offshore showing the sea bottom, superficial deposits, rockhead and two bedrock formations. Vertical scale is in milliseconds. (Courtesy of Huntec Ltd.)

symmetrical and may appear offset from the position of the actual feature being explored.

Gravitational maps which trace anomalies due to variation in rock densities (known as *Bouguer anomalies*) have the advantage that the anomalies appear directly over features, but also suffer from the disadvantage of only picking up large changes in density. The method has limited application in civil engineering due to this scale problem.

Seismic refraction and reflection surveys produce sections which can be very useful in tracing rockhead (Figure 3.10) and other underground structures to various depths depending on the size of the impulse and the spread of the geophones. Seismic reflection profiles used over water (Figure 4.17) can be particularly advantageous in coastal and offshore work.

Resistivity and ground conductivity contour maps are normally used for moderate to shallow depths to determine the position of buried channels, landfill, gravel deposits or old mine shafts (see Figure 4.18). The more recent ground probing radar, produces a profile to shallower depths which may show hollows, voids, pipes or even depth of soft or loose superficial deposits (see Figure 4.19).

4.4.5 *Hydrogeological maps*

Regional maps. Details of groundwater and surface water aspects of an area are often shown on engineering geological maps. European multi-purpose maps have hydrogeological features added in blue. These normally include *hydroisobaths* (contours on the maximum seasonal level of the groundwater table) together with specific symbols (see Appendix B) for corrosiveness (including pH, hardness and

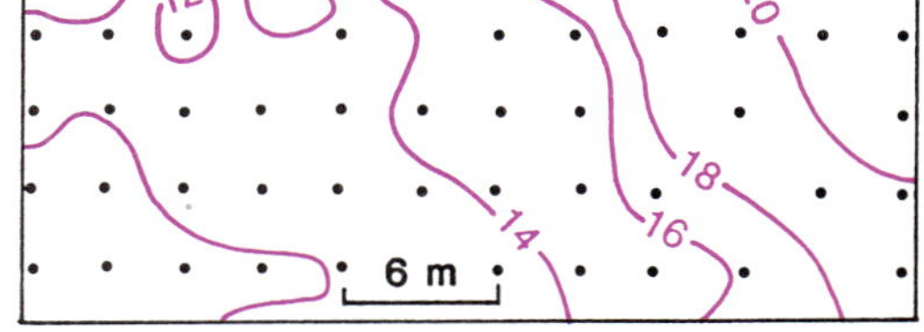

Figure 4.18 Ground conductivity survey over buried mineshaft. Contours in mmhos m^{-1}. • Grid of stations at 3 m intervals. (Alta Geophysics, Birmingham.)

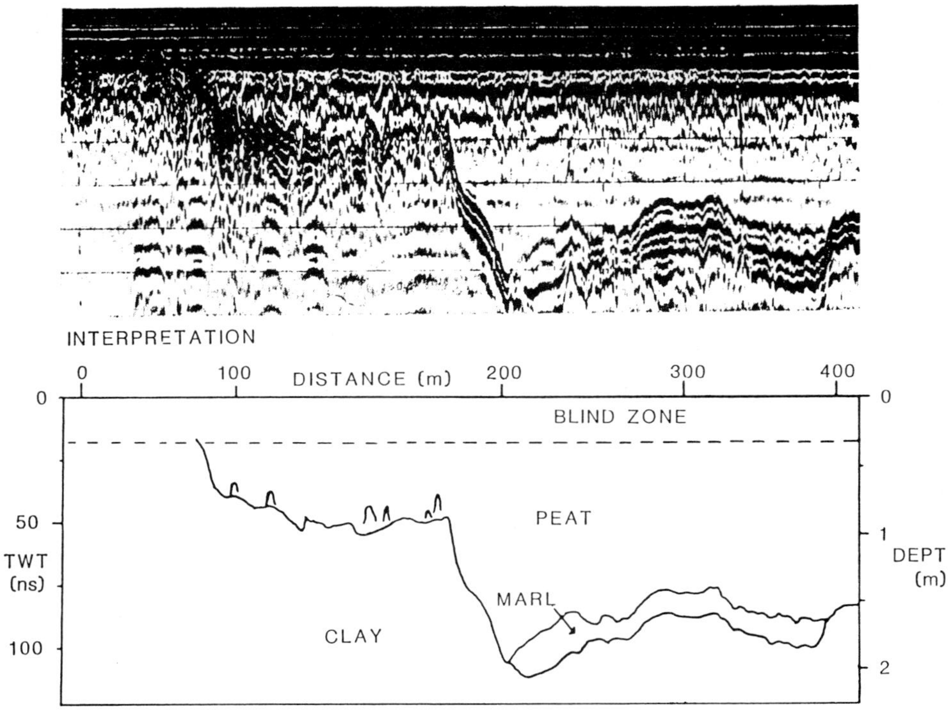

Figure 4.19 Ground probing radar survey over an area of peat overburden in Ireland. TWT is the two way travel time, in nanoseconds. (From Working Party Report.[19])

aggressivity of CO_2 and SO_4). Surface features such as springs, artesian conditions, and waterlogged ground are also given, even on small-scale maps.

Whilst there appears to be less emphasis on groundwater conditions on engineering geology maps in Britain, the British Geological Survey have published a number of hydrogeological maps covering appropriate parts of the country. These started in the south and east on scales of 1:63 360 and 1:126 720 but in recent years these maps have become standardized at 1:100 000. Since the maps are made for hydrogeological purposes they contain detailed information about water bearing rocks (*aquifers*) and sometimes have supplementary insets showing groundwater conditions for individual formations with numerous structure contours and isopachytes as well as water table contours. Details of fluoride, chloride and total hardness concentration zoning may also be included. They also give surface features but are mostly designed for groundwater extraction purposes or where groundwater aquifers are of major importance to a project (Plate 4*b*).

Groundwater on site. For most construction purposes, groundwater data are normally more site-specific. Boreholes may be used to determine water levels, permeability values, sulphate contents and pore water pressures, which then can be most conveniently represented on sections through the site. From these sections, it may be possible to assess water ingress or dewatering procedures where groundwater control is necessary (see Figure 4.10).

Groundwater pollution. Pollution of groundwater from various sources is an important aspect of the hydrogeological mapping of aquifers especially in urban areas or near landfill sites. Again, boreholes or test wells may supply the required data on water chemistry and movement, but one way in which polluted or saline waters can be mapped in homogeneous aquifers is by using electrical resistivity or conductivity values. For example in chalk, the resistivity varies from less than 20 ohm/m for saline groundwater to levels well over 100 ohm/m for normal groundwaters.

One resistivity map from the Midlands of England (Figure 4.20) shows how low values of resistivity in a Triassic sandstone aquifer delineated an area of highly saline groundwater and helped to define the position of the bounding Tixall fault. It explained why only one out of four pumped boreholes contained saline groundwater, presumably derived from salt-rich Triassic units. The 58 resistivity soundings proved very much more cost-efficient than numerous observation wells.

Waste disposal is a critical issue both where shallow sites are involved for domestic, industrial and low level radioactive wastes, and where deep storage is concerned with higher level wastes.

Shallow sites can build up toxic concentrations or *leachates*, which must be collected or dispersed before reaching a groundwater system, since this may affect river drainage or any aquifer supply. Most modern landfill sites are therefore controlled by multiple barriers, cut-off trenches and impermeable cappings. Detailed plans and sections giving the design procedures and operational practice on sites must be closely adhered to in order to assure the necessary quality control. Subsequent monitoring of both groundwater quality and flow should be accompanied by similar concern for the build up of gases, particularly methane, which can sometimes be collected in a similar manner to leachate (Figure 4.21).

The thorny question of deep disposal of more

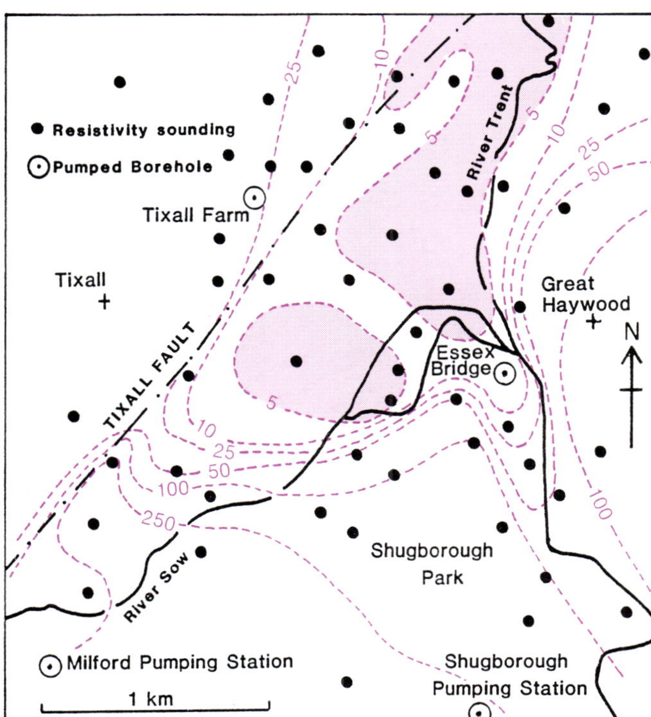

Figure 4.20 Resistivity contours in ohm-metres taken from 58 resistivity soundings in central England showing the concentration of saline groundwater north and west of Essex Bridge borehole. (After Barker, 1986.[20])

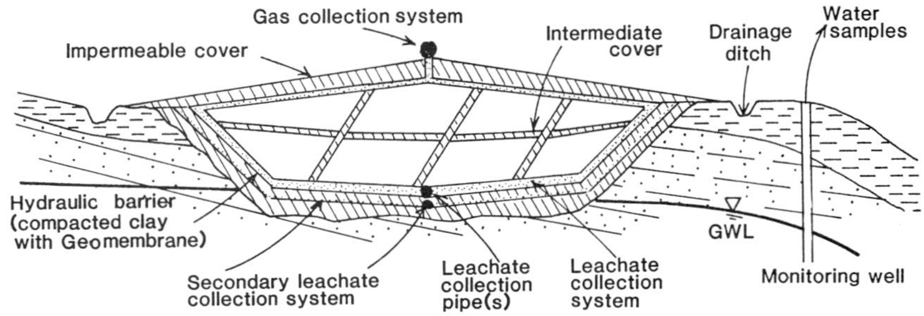

Figure 4.21 Typical cross-section of a modern, below-ground landfill site with leachate and gas collection systems and a double liner.

active chemical and radioactive wastes is being tackled in many countries, notably Sweden, Germany, Canada and the United States, with some urgency. Site investigations and detailed geological considerations are of paramount importance so that the rock mass characteristics, associated structure and mineralogical compositions are considered in terms of both excavation techniques and possible migration of contaminants. For these types of projects, involving complex galleries sometimes at great depths (Figure 4.22), the most sophisticated methods of numerical modelling and the plotting of underground structures in three dimensions are

Figure 4.22 Typical complex layout of underground chambers in crystalline rock associated with an offshore deep repository, Sweden. (After Morfeldt, 1989.[26])

essential. Deep boreholes with sampling, geophysical logging and pilot schemes may be necessary before engineers can be certain that no faults, shears or other fractures exist or are likely to develop in the rocks at depth.

One of the materials favoured in Germany is rock salt within underground salt domes. The ductility of the salt is favourable to lack of fracturing, but has the disadvantage of being susceptible to the convergence of an opening with time. Argillaceous or crystalline rocks of low permeability are also favoured by geologists for their physical, hydrogeological and geochemical properties, all of which must be studied in great detail before the choice of site is finalised.

Plates

Plate 1 Extracts from (*a*) Drift and (*b*) Solid editions of part Stirling sheet (Scotland 39). Line of section is in black (see Figure 3.5) and line of motorway is in red (see Question 3, Appendix A). Colours and symbols are explained below. (Reproduced at a scale of 1:63 360 with the permission of the Director, British Geological Survey. NERC/Crown copyright reserved.)

(*a*) Drift edition

	Colour	*Symbol*
Reclaimed intertidal flats		R
Recent river alluvium	Yellow	
Peat	Pale brown	
Post-glacial estuarine deposits		PG
Late-glacial estuarine deposits		LG
Former coastline	Black with brown, margin downslope	
Glacial sand and gravel	Pink	
Glacial till	Blue	
Rock at or near surface	Grey	
Drift and rockhead boundaries		——·····

(*b*) Solid edition

		Colour	*Symbol*
Carboniferous	Limestone coal group		d^{M1}
	Lower limestone group		d^{L4}
	Calciferous sandstone measures (with volcanic detritus)		$d^{L1-3}_{\text{(with red dots)}}$
	Basaltic lavas of various types		B, B^{M}, B^{J}, Mi^{B}
	Lower calciferous sandstone measure with sandstone	Green–grey	d^{LI-3}
Devonian	Cornstone beds		C^{3C}
	Upper Old red sandstone		C^{3}
	Pyroxene andesite lavas		pA
	Tuff		Z
	Basaltic lava		fB
	Volcanic conglomerate		$C^{1}_{\text{(with red dots)}}$
	Lower ORS Sherrifmuir Formation		C^{1S}
	Lower ORS Cromlix Formation		C^{1C}
	Quartz dolerite		Q^{D}
	Dolerite		D, K
	Top Hosie limestone		Top.Ho
	Murrayshall limestone		Mu.
	Faults with crossmarked downthrow side	Red	—┬—

Plate 1

a.

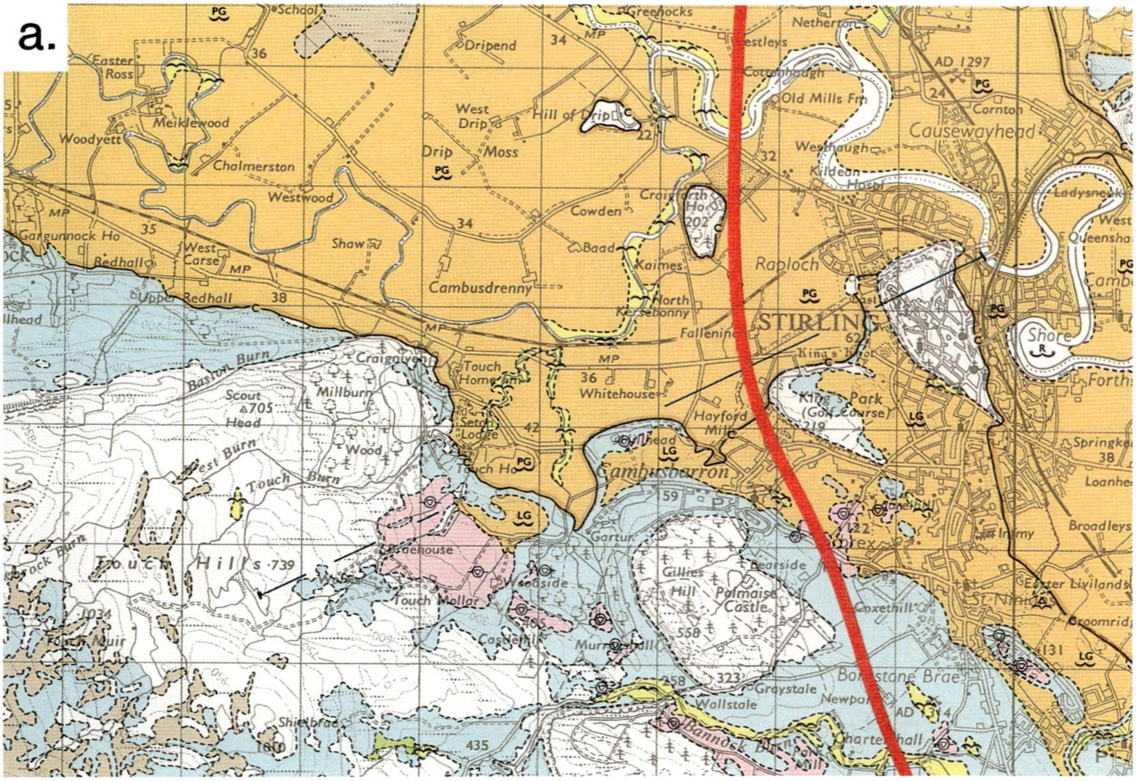

School
Greenocks
Dripend
Netherton
Easter Ross
Meiklewood
Woodyett
Chalmerston
West Drip
Hill of Drip
Drip Moss
Cottonhaugh
Old Mills Fm
Causewayhead
Cornton
AD 1297
Westhaugh
Kildean Hospl
Ladysneuk
Westwood
Gargunnock Ho
Redhall
West Carse
Upper Redhall
Shaw
Cowden
Baad
Kaimes
Cambusdrenny
North Kersebonny
Raploch
STIRLING
Fallenin
King's Knot
King's Park (Golf Course)
Shore
Forth
Baston Burn
Craigniven
Scout Head
Millburn
Wood
Touch Home Fm
Seton Lodge
Whitehouse
Touch Ho
West Burn
Touch Burn
Hayford Mill
Hillhead
Cambusbarron
Springkerse
Loanhead
Broadleys
Touch Hills
Gartur
Gillies Hill
Polmaise Castle
Bearside
Woodside
Touch Mollar
Castlehill
Murrayshall
Coxethill
Easter Livilands
St Ninians
Broomridge
Touch Muir
Borestone Brae
Graystale
Wallstale
Newpark
Chartershall
Shielbrae
Bannock Burn
AD 1314

b.

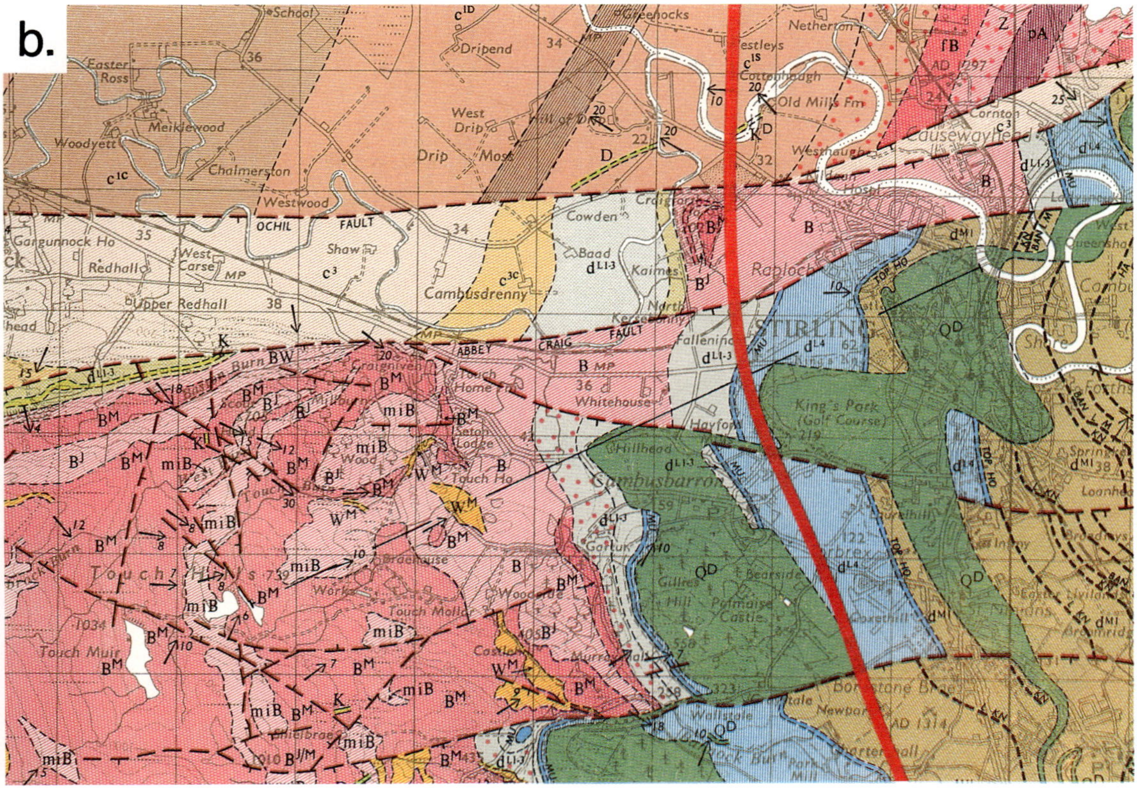

School
Greenocks
Dripend
Netherton
Easter Ross
Meiklewood
Woodyett
Chalmerston
West Drip
Drip Moss
Old Mills Fm
Causewayhead
Cornton
Westwood
OCHIL FAULT
Gargunnock Ho
Redhall
West Carse
Upper Redhall
Shaw
Cowden
Baad
Kaimes
Cambusdrenny
Raploch
ABBEY CRAIG FAULT
STIRLING
King's Park (Golf Course)
Hayford
Whitehouse
Touch Ho
Cambusbarron
Hillhead
Gartur
Gillies Hill
Polmaise Castle
Bearside
Woodside
Touch Mollar
Coxethill
St Ninians
Touch Hills
Touch Muir
Shielbrae
Borestone Brae
Newpark
AD 1314
Chartershall

Plate 2 Extracts from Solid and Drift editions Scotland sheet 15 of the Sanquhar area of Southern Scotland. These illustrate the clarity of the revised 1:50 000 edition (*c* and *d*) compared with the original 1870 solid sheet (*a*) and the 1937 Drift edition (*b*). (Reproduced with the permission of the Director, British Geological Survey. NERC/Crown copyright reserved.)

(*a*) 1:63 360 First edition (1870)

N.B. These early maps were hand-coloured solid editions with the addition of alluvial deposits. No contours.

	Colour	*Symbol*
Alluvium	Yellow	
Coal Measures	Dark grey, seams in black	
Lower Carboniferous	Blue–grey	d^1
Old Red Sandstone (ORS conglomerate)	Red–brown (with red dots)	
ORS lavas	Pink	
Lower Palaeozoic with coarse bands	Pale grey	
	Yellow dots	
Intrusive Granodiorite		G
Dolerite	Red	
Felsite	Orange	F
Faults	White	

Other symbols as in Figure 1.1*b*

(*b*) 1:63 360 Drift edition (1937)

This map shows a variety of drift materials at or near the surface and includes various rock types with symbols as in Figure 1.1*b* other than:

d^5 for d^{C1} of Figure 1.1*b*
d^4 for d^M " "
and Sb represents an Antimony vein

(*c*) 1:50 000 Solid edition (1986)

Recent compilation sheet based on remapping 1957–1960. Symbols as in Figure 1.1*b*

(*d*) 1:50 000 Drift edition (1982)

Compilation of drift deposits taken from 1:10 560 sheets but using colours without symbols.

	Colour
Man-made fills	Pale green
Peat	Brown
River alluvium	Yellow
Glacial meltwater deposits	Pink
Glacial moraine (ablation till)	Green
Glacial boulderclay (lodgement till)	Blue
Bedrock	Mauve

Compare the glacial drainage channel symbol (thick arrow with tail) with that used on older drift map and 1:10 560 map (Figure 1.1)

Plate 2

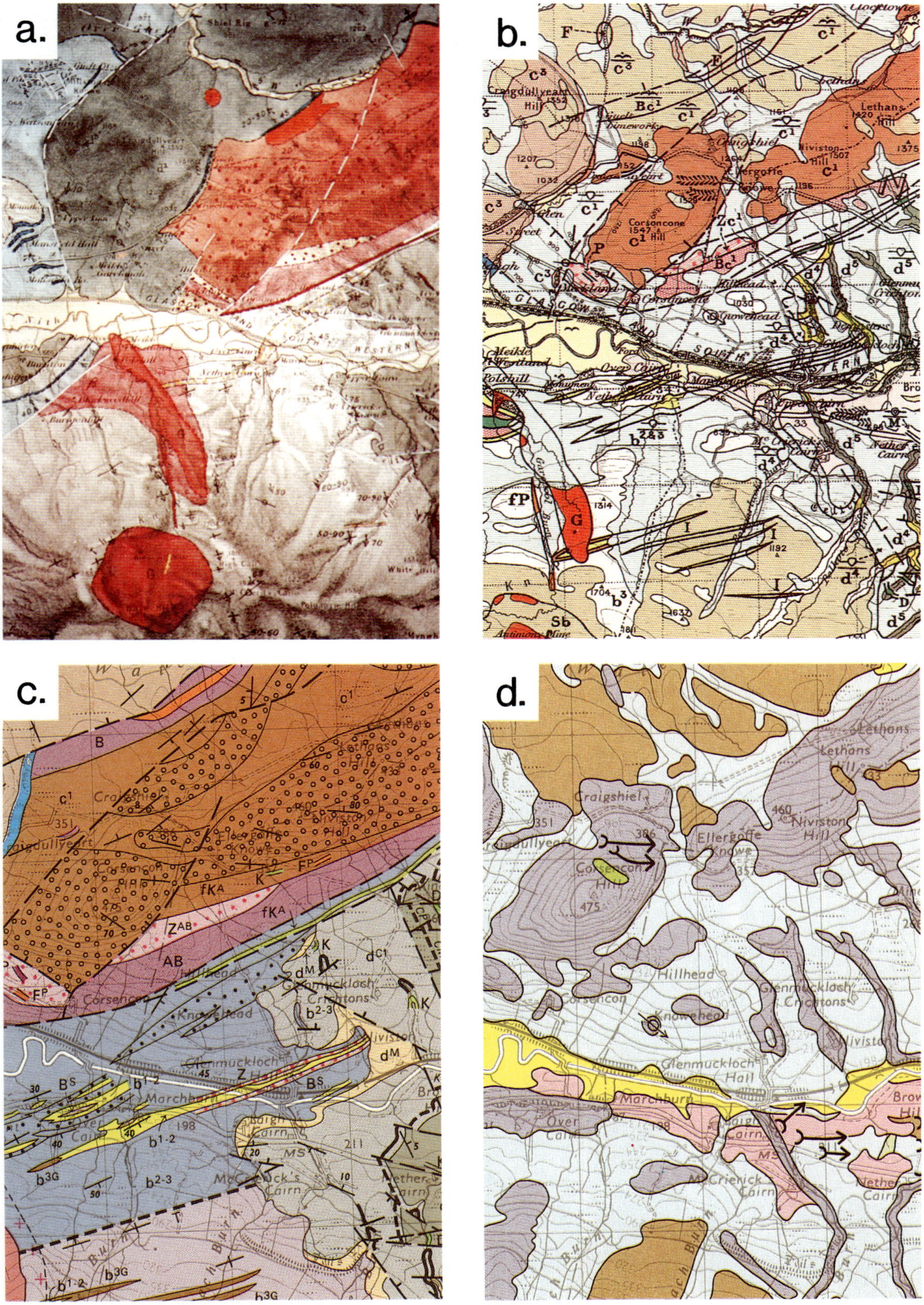
a.
b.
c.
d.
Lethans Hill
Niviston Hill
Corsoncone Hill
Craigdullyeart Hill
Glenmuckloch
Marchburn
Knowehead
Hillhead
Craigshiel
Ellergoffe Knowe
Glenmuckloch Hall
Over Cairn
Nether Cairn
McCrierick's Cairn
Antimony Mine
Glenmuckloch Craigtons

Plate 3 Extracts from four engineering geology maps of the upper Forth estuary published by the Engineering Geology Unit of the British Geological Survey at a scale of 1:50 000.[2]

(*a*) Sheet 6: Engineering classification of surface sediments

A drift sheet with re-classification of materials in engineering terms as follows:

	Colour
Landfill	Green
Peat	Pink
Recent intertidal soils	Deep yellow
Normally consolidated cohesive soils	Pale yellow
Cohesive soils with desiccated crust	Orange
Cohesionless soils	Brown
Over-consolidated cohesive soils	Blue
Bedrock	Mauve
Former coastline	Thick, curved black line

(*b*) Sheet 2: Drift thickness contours (depth to rockhead)

Contours show depth to rockhead in metres. Lines are dashed to indicate uncertainty, and shallow rockhead areas are coloured: <3 m, blue; at or near surface, green. Some engineers may still prefer rockhead contours reduced from sea level, but since this area is so flat the two are very similar.

(*c*) Sheet 1: Engineering geology of the solid rocks

Colours are main geological groups.

	Colour
Millstone Grit	Yellow
Coal measures	Orange

In addition:

	Symbol
Areas of thin *glacial* drift	Horizontal lines
Areas where rock within 3 m of surface	Dark tone
Areas of suspected shallow mine workings	Diagonal line
Location of test borehole	●14
Location of rock mass assessments with classification (I–V) indicated	▲ II
Major faults	Dashed lines

(*d*) Sheet 8: Geotechnical planning map for heavy structures

This extract summarises the ground conditions for heavy loading from A to F, colour-coded blue to pink. The zones range from very good to very poor and are sub-divided into material classes (1–21) as in the following examples:

$\mathbf{C}\frac{18,19}{2,4,5}/\mathbf{I}$ = Fair foundation site of soft–very soft clays and silts over various sedimentary rocks/reliable zone assessment (**I**)

$\mathbf{A_2/O}$ = Very good foundation site with strong limestones and sandstones/based on limited geotechnical data (**O**).

Plate 3

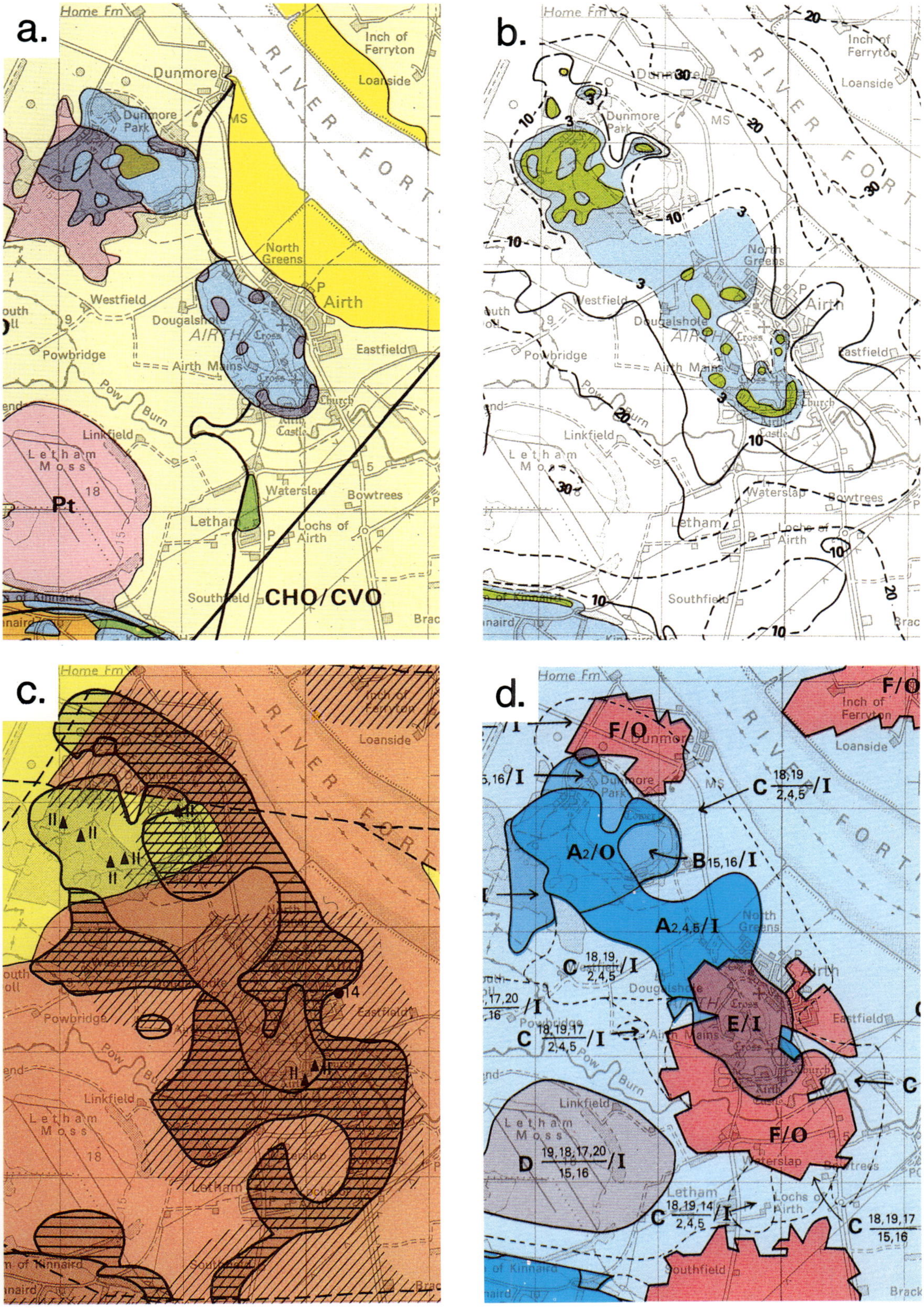
a.
Home Fm
Inch of Ferryton
Loanside
Dunmore
RIVER FORTH
Dunmore Park
MS
North Greens
Westfield
Airth
Dougalshole
AIRTH
Cross
Eastfield
Powbridge
Airth Mains
Pow Burn
Church
Airth Castle
Linkfield
Letham Moss
18
Pt
Waterslap
Bowtrees
Letham
Lochs of Airth
Southfield
CHO/CVO
b.
20
30
10
3
c.
14
d.
F/O
C 18,19/2,4,5 /I
A2/O
B15,16/I
A2,4,5/I
C 18,19/2,4,5 /I
E/I
C 18,19,17/2,4,5 /I
D 19,18,17,20/15,16 /I
C 18,19,14/2,4,5 /I
C 18,19,17/15,16

Plate 4

(*a*) Map showing relative ease of excavation, Utah, USA. Extract from USGS Folio of the Salina Quadrangle, Utah, Map I-591-J, 1972. Scale approximately 1:170 000. Excavation categories 1–5 are assigned as follows: 1 Excavation very easy; 2 Excavation easy; 3 Excavation easy–difficult due to interbedding; 4 Excavation difficult; 5 Excavation very difficult. 1–3 can be removed by hand or light machine, whereas 3–5 may require heavier plant or blasting. (Reproduced after part of the map by P. L. Williams, United States Geological Survey. Redrawn by UNESCO/IAEG, 1976.[6])

(*b*) Extract from BGS Hydrogeological map of the area south of the Dartford district of Kent, England, at a scale of 1:63 360.

Stratigraphy

		Colour
Tertiary Period	London clay (aquiclude)	Blank
	Blackheath beds (sands)	Red dots
	Woolwich beds (aquiclude)	Blank
	Thanet beds (sands)	Blue dots
Cretaceous Period	Upper chalk	Green brick

Symbols

Structure contours on base of Middle chalk in feet	Green
Structure contours on top of Upper chalk in feet	Blue
Hydro-isobaths on chalk water table (1966)	Purple
Surface water divide	Black dots
Groundwater divide	Purple circles
Springs	Blue circles with arrows
Water wells and boreholes in chalk	Red symbols

Plate 4

a.

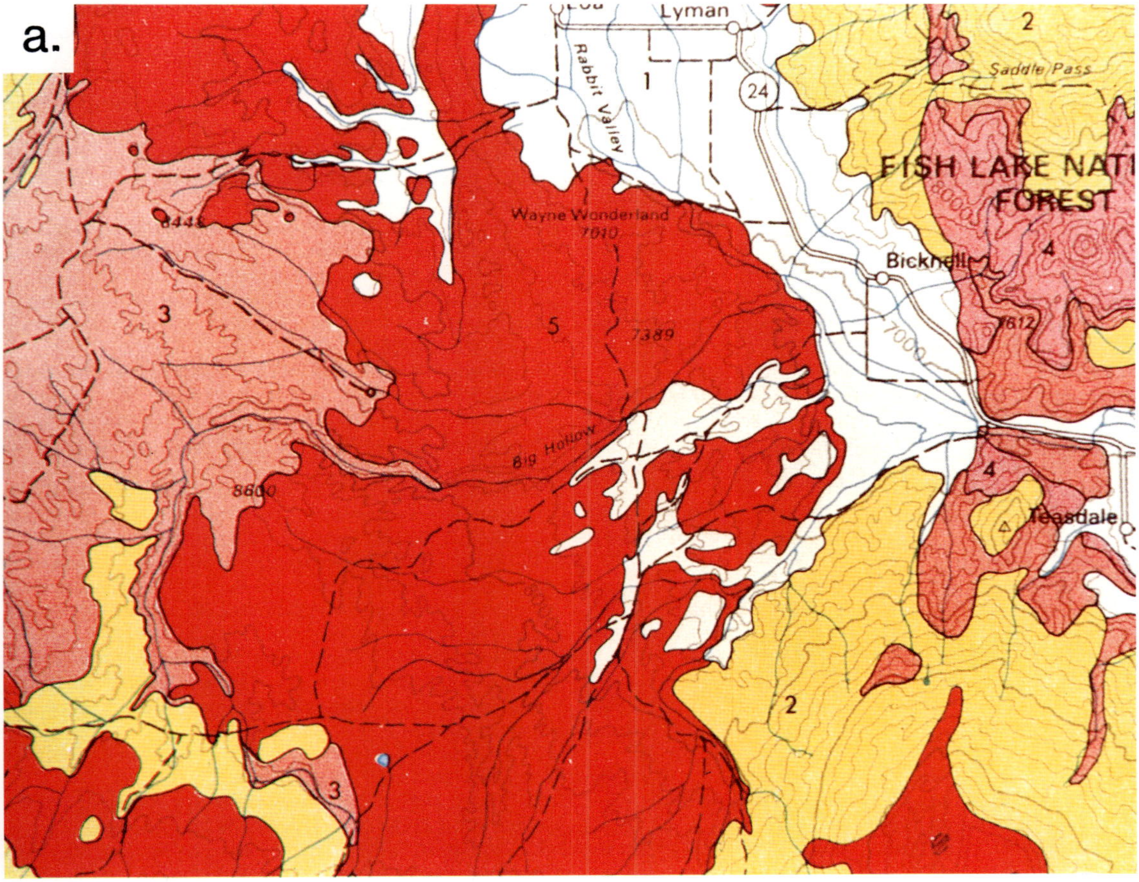
Lyman
Rabbit Valley
24
Saddle Pass
FISH LAKE NAT
FOREST
Wayne Wonderland
7010
Bicknell
7000
7812
7389
8448
Big Hollow
8600
Teasdale
1
2
3
4
5

b.

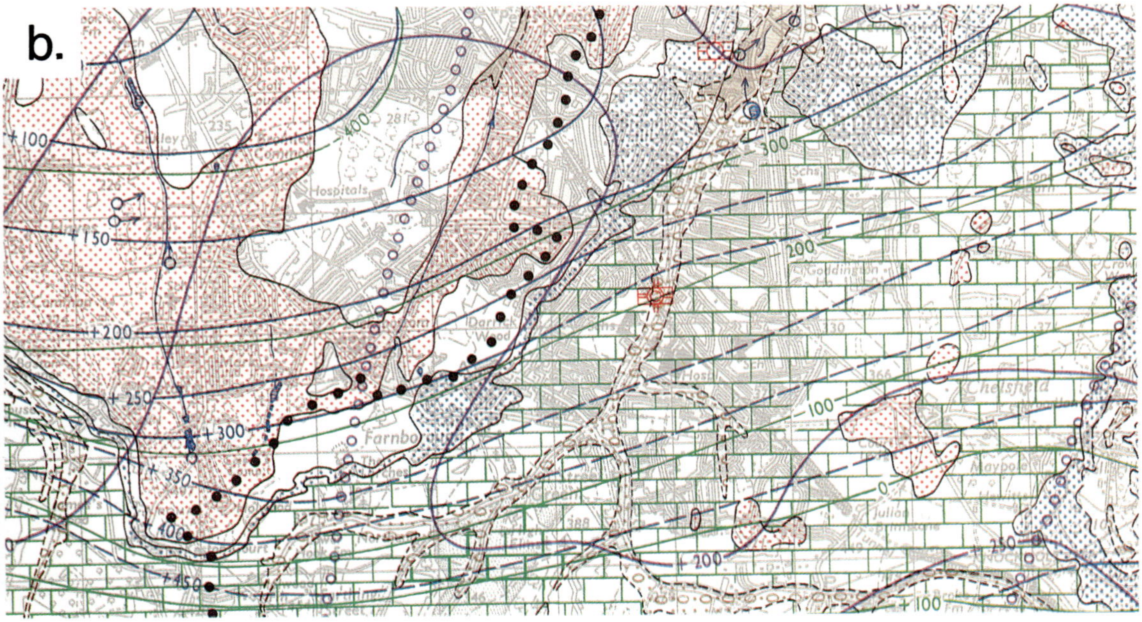
Hospitals
Farnbo
Chelsfield
Maypole
Goddington
Derrick Wood
+100
+150
+200
+250
+300
+350
+400
+450
400
300
200
100
0
+200
250
+100

References

1. Lenssen, S. (1973) The story of Borehole 4, Kariba North. *New Civil Engineer* 8th February, 10.
2. Gostelow, T.P. and Browne, M.A.E. (1976) Engineering geology of the upper Forth Estuary. *Rep. Br. Geol. Surv.* **16**, No. 8
3. Sheppard, T. (1920) *William Smith: His Maps and his Memoirs*. Brown & Sons, Hull.
4. Anon (1972) The preparation of maps and plans in terms of engineering geology. Working Party Report. *Q. J. Engng. Geol.* **5**, 293–381.
5. Dearman, W.R. and Fookes, P.G. (1974) Engineering geological mapping for civil engineering practice in the United Kingdom. *Q. J. Engng. Geol.* **7**, 223–256.
6. UNESCO/IAEG (1976) *Engineering Geological Maps. A Guide to their preparation.* The UNESCO Press, Paris.
7. Forsyth, I.H., McMillan, A.A., Browne, M.A.E. and Ball, D.F. (1985) Account accompanying environmental geology maps of Glasgow. BGS, Open File report.
8. Brink, A., Partridge, T. and Williams, A. (1974) *Soil Survey for Engineering*. Oxford University Press, Oxford.
9. Anon (1977) The description of rock masses for engineering purposes. Working Party Report. *Q. J. Engng. Geol.* **10**, 355–388.
10. Hawkins, A.B. (1986) Rock descriptions. In: *Site Investigation Practice. Assessing BS.5930*, Hawkins, A.B. (Ed.). Geol. Soc. Lond. Engng. Geol. Spec. Publ. No. 2, 59–84.
11. Cottiss, G.I., Dowell, R.W. and Franklin, J.A. (1971) A rock classification system applied in civil engineering. *Civil Engng. Public Works Rev.* **611**, 736–738.
12. Anon (1981) BS 5390 (Site investigations). British Standards Institution.
13. Knill, J.L. and Jones, K.S. (1965) The recording and interpretation of geological conditions in the foundations of the Roseires, Kariba and Latiyan dams. *Geotechnique* **15**, 94–124.
14. Wickham, G.E., Tiedemann, H.R. and Skinner, E.H. (1972) Support determination based on geological predictions. In: *Proc. 1st N. Am. Tunneling Conf.* A.I.M.E., New York, 43–69.
15. Bieniawski, Z.T. (1974) Geomechanics classification of rock masses and its application in tunneling. In: *Proc. 3rd Int. Cong. Rock Mech.*, Denver, **2**, 27–32.
16. Bieniawski, Z.T. (1976) Rock mass classification in rock engineering. In: *Proc. Symp. Exploration for Rock Engineering*, AA. Balkema, Cape Town, **1**, 97–106.
17. Barton, N. (1976) Suggested methods for quantitative description of discontinuities in rock masses. *Int. J. Rock. Mech. Min. Sci.* **15**, 319–368.
18. Higginbottom, I.E. (1976) The use of geophysical methods in engineering geology. *Ground Engng.* **9**, 34–38.
19. Anon (1988) Engineering geophysics. Report by the Geol. Soc. Engng. Group Working Party. *Q. J. Engng. Geol.* **21**, 207–271.
20. Barker, R.D. (1986) Surface geophysical techniques. In: *Groundwater: Occurrence, Development and Protection*, Brandon, T.W. (Ed.). Inst. Water Eng. and Scientists, London.
21. Anon (1981) Recommended symbols for engineering geological mapping. Report by the IAEG Commission on engineering geological mapping. *Bull. Int. Assoc. Eng. Geol.* **24**, 227–234.
22. Hoek, E. (1987) General two-dimensional slope stability analysis. In: *Analytical and Computational Methods in Engineering Rock Mechanics*, Brown, E.T. (Ed.). Allen and Unwin, London.
23. Phillips, F.C. (1971) *The Use of Stereographic Projection in Structural Geology*. Edward Arnold, London.
24. Ragan, D.M. (1985) *Structural Geology. An Introduction to Geometrical Techniques* (3rd edn.). Wiley, New York.
25. Woodland, A.W. (1968). Field geology and the civil engineer. *Proc. Yorks. Geol. Soc.* **36**, 531–578.
26. Morfeldt, C.O. (1989) Different subsurface facilities for geological disposal of radioactive waste (storage cycle) in Sweden. *Bull. Int. Assoc. Eng. Geol.* **39**, 25–33.
27. Paul, M. A. and Balfour, J. A. D. (1990) Computer-assisted teaching of geological map interpretation to undergraduate civil engineers. *Proc. Instn. Civ. Engrs.* Part 1 June 367–380.

Appendix A Map exercises

Engineering geological problems

These questions are based on figures included in this book. For outline solutions see Appendix E.

(1) *Figure 1.1*: Consider the geological problems associated with a proposal for a dam crossing the main valley at a top water level of 700 ft and following the Easting Grid Line (shown by the thick red line).

(2) *Figure 1.1*: If the dam in (1) was not built, what geological problems might affect the proposed site of a factory complex approximately 500 sq m, north of the main road in the SE corner of the sheet (outlined by thin red line)?

(3) *Figure 1.1*: Predict geological conditions which may be expected in a sewer tunnel running at a constant gradient from the NE corner of the area (shaft invert at 850 ft to a junction at Portland Place in the west (invert 580 ft OD) (shown as twin red lines)?

(4) *Figure 1.2*: Using other information to find the position of places mentioned below, what quality of map would you expect to find for sites at (i) Cardigan, Wales (ii) Southern Loch Ness, Scotland, (iii) Morecambe, Lancashire, (iv) The Isle of Wight, (v) Dundee, Scotland, (vi) Whitby, Yorkshire, (vii) Hastings, Sussex and (viii) Orkney, Scotland?

(5) *Plate 1*: Draw a section along the line of the M9 motorway west of Stirling (marked in red) and write a report on recommended site investigation procedures, for the whole section shown on the map.

(6) *Plate 2*: Study the two Solid and two Drift editions of the same area shown in Plate 2 and discuss improvements which have been achieved in each.

(7) *Figure 4.6*: Write brief comments on the problems likely to be associated with the geology of this tunnel.

(8) *Plate 3*: Using the plate explanation write an engineering geological report describing the ground conditions in the area shown in the extracts.

(9) *Plate 4a*: The area between Teasdale and Bicknell on the extract from the Salina Quadrangle, USA, may be considered to be a potential site for a large dam. Comment on the ease of excavation of foundations with a top water level of 8000 ft. What further problems may be indicated by the classification system used on the map?

Problem maps and plans

The following maps and plans (Problem Maps A.1–A.7) have been designed to allow the reader to put into practice some of the methods outlined in Chapters 2 and 3. They are geometrically simplified to allow for completion in about 30 min, although some of the later ones may take up to an hour to complete both the calculations and a description. Whilst the author appreciates that outline maps of this type are no substitute for some of the real maps shown in this book, they do help to illustrate particular points of principle, which should help the engineer to deal with maps more confidently.

Problem Map A.1a

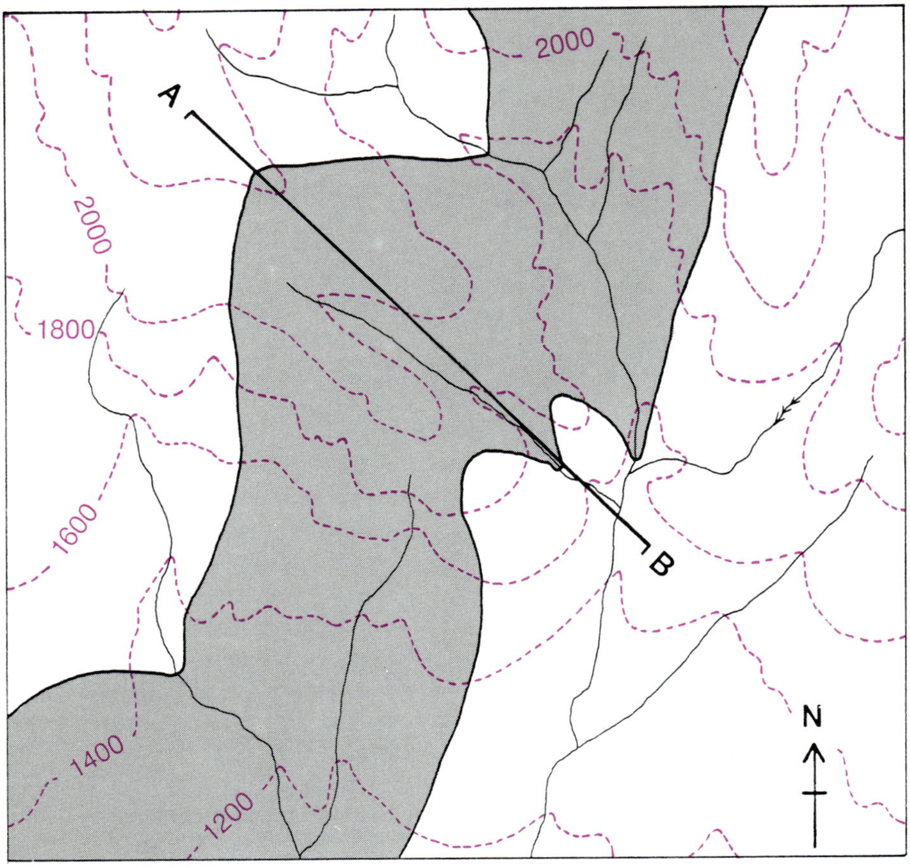

Scale 1:50,000

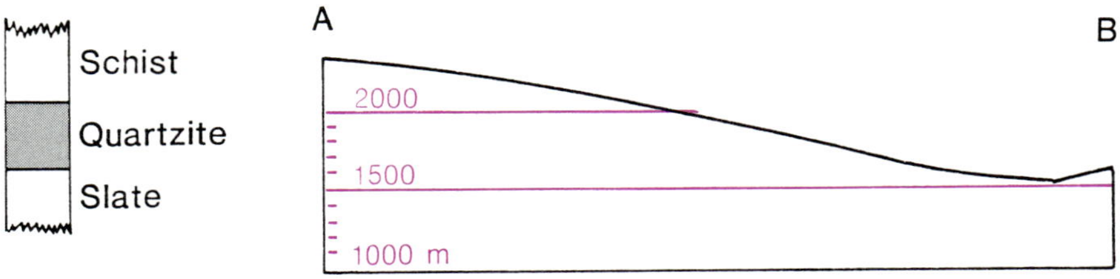

Problem Map A.1a Simple dip.
Problem: (i) Construct structure contours for the base of the quartzite. (ii) Calculate the constantly dipping angle of true dip, α, and its azimuth, β. (iii) Plot the quartzite on the section AB.

Problem Map A.1b

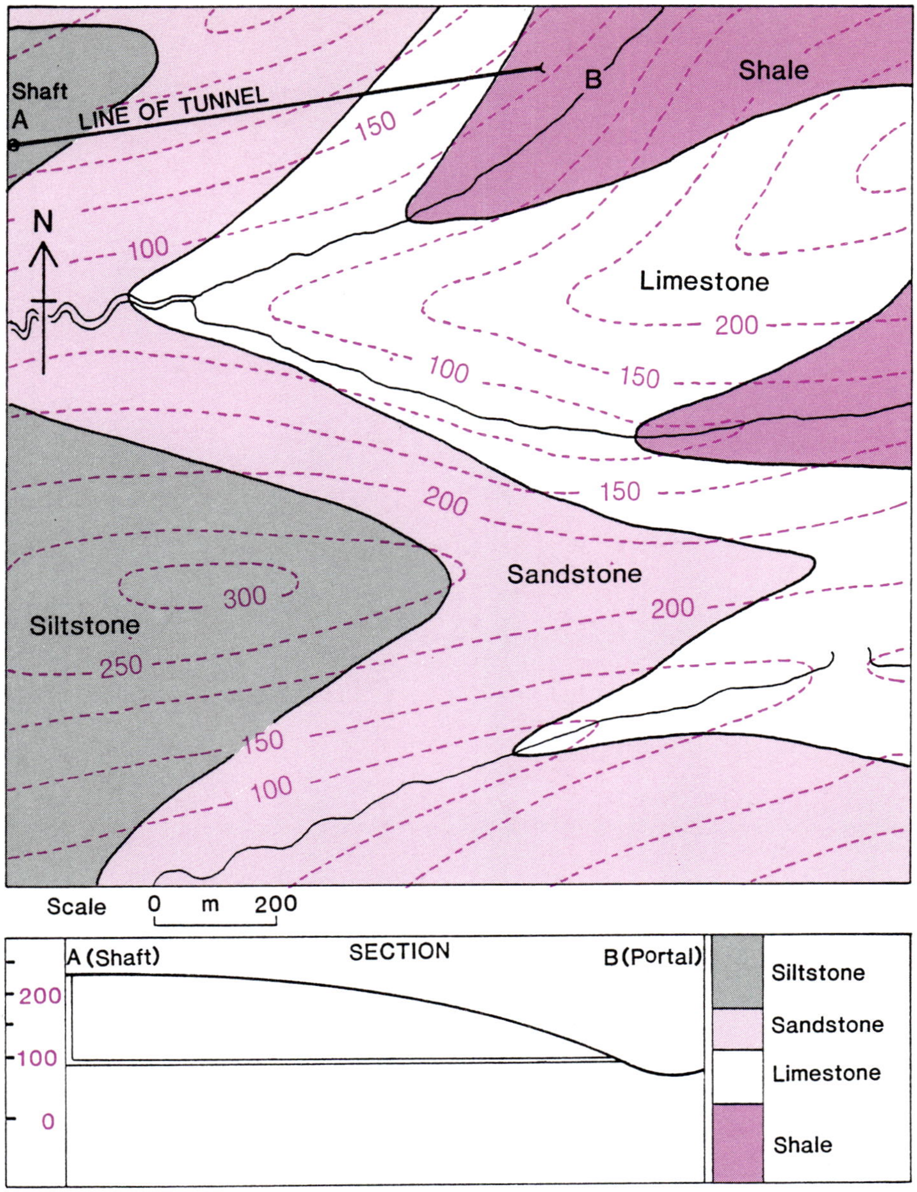

Problem Map A.1b Strata dipping in a tunnel.

Problem: (i) Calculate the angle and direction of dip. (ii) Estimate the vertical thicknesses, where possible. (iii) Plot the geology of the tunnel line AB (Note: horizontal invert is at 95 m above sea level). (iv) Calculate chainages from B for geological boundaries, and give the depth to sandstone in shaft A. (v) Measure apparent dip in the tunnel from the tunnel section AB (Note: section is true scale).

Problem Map A.2a

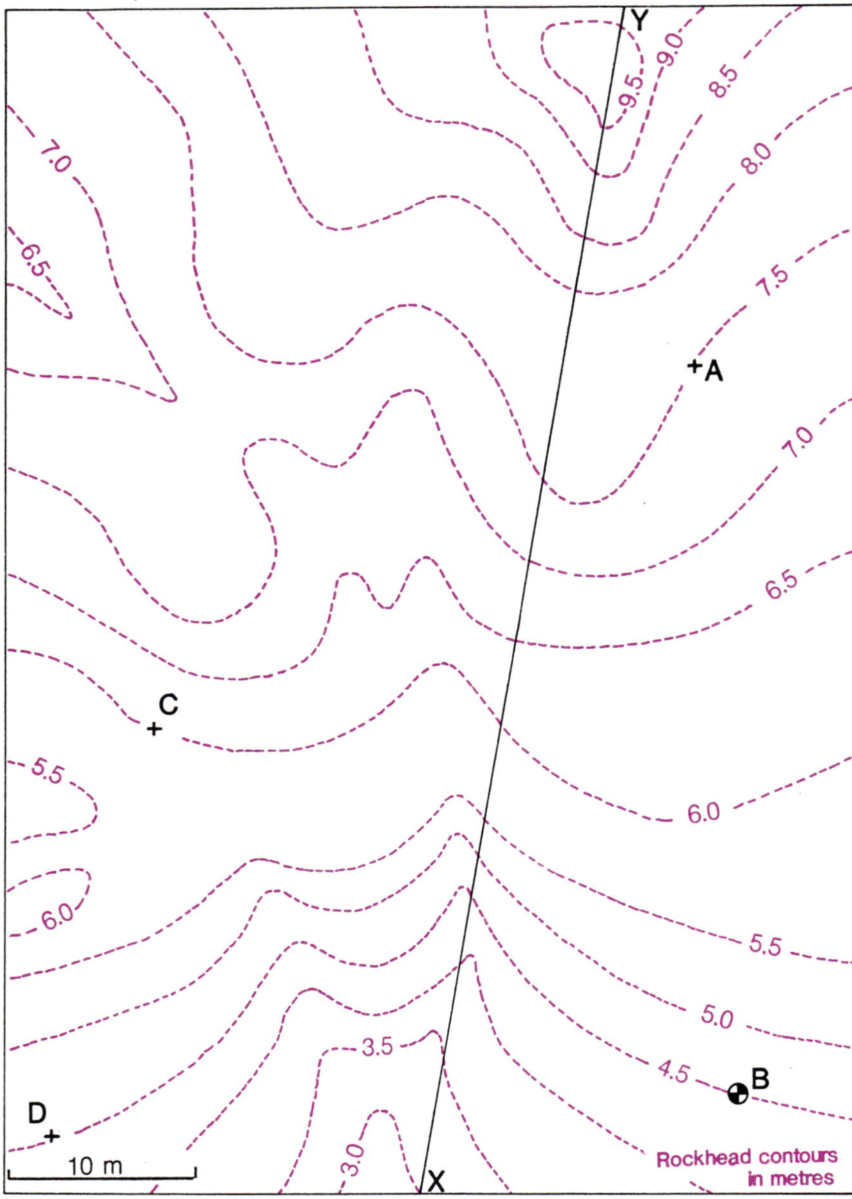

Problem Map A.2a Three-point solution.

At the foundation site, borehole B proved the downwards succession below rockhead to be: mudstone, 1.5 m; coal, 0.5 m; seatearth, 2 m; and sandstone, at least 1.5 m. The base of the mudstone outcrops at point C, and the base of the coal outcrops at point A. Contours are rockhead contours in metres.

Problem: (i) Plot the predicted outcrop of the beds at rockhead across the site, and calculate the angle and direction of dip (which may be assumed to be constant). (ii) What is the depth of the coal at D? (iii) Shade the area where the coal seam does *not* underlie the site. (iv) On a separate sheet, construct a geological section along the line XY.

Problem Map A.2b

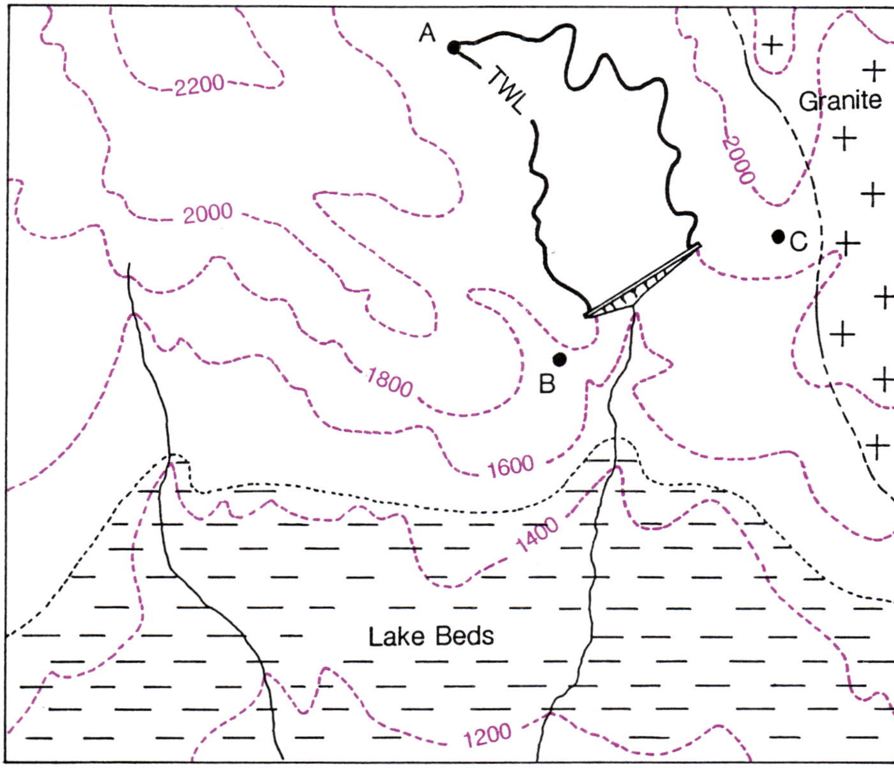

Scale 1:50,000

Summary logs for boreholes A, B and C

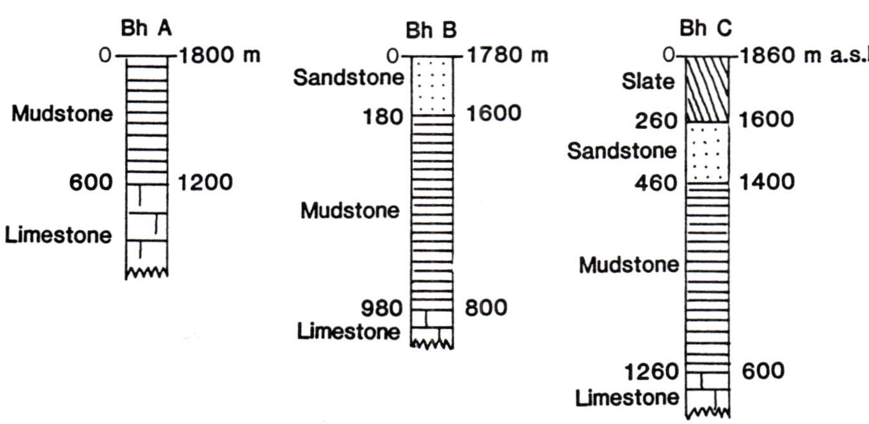

Problem Map A.2b Reservoir problems.
The sedimentary section (shown blank on the map) is very poorly exposed in heavily wooded hill country. In the few exposures found, it appears to have the same dip.
Problem: Use the three boreholes available to plot the predicted boundaries and then discuss possible reservoir problems associated with the proposed dam site, which gives a top water level of 1800 m.

Problem Map A.3

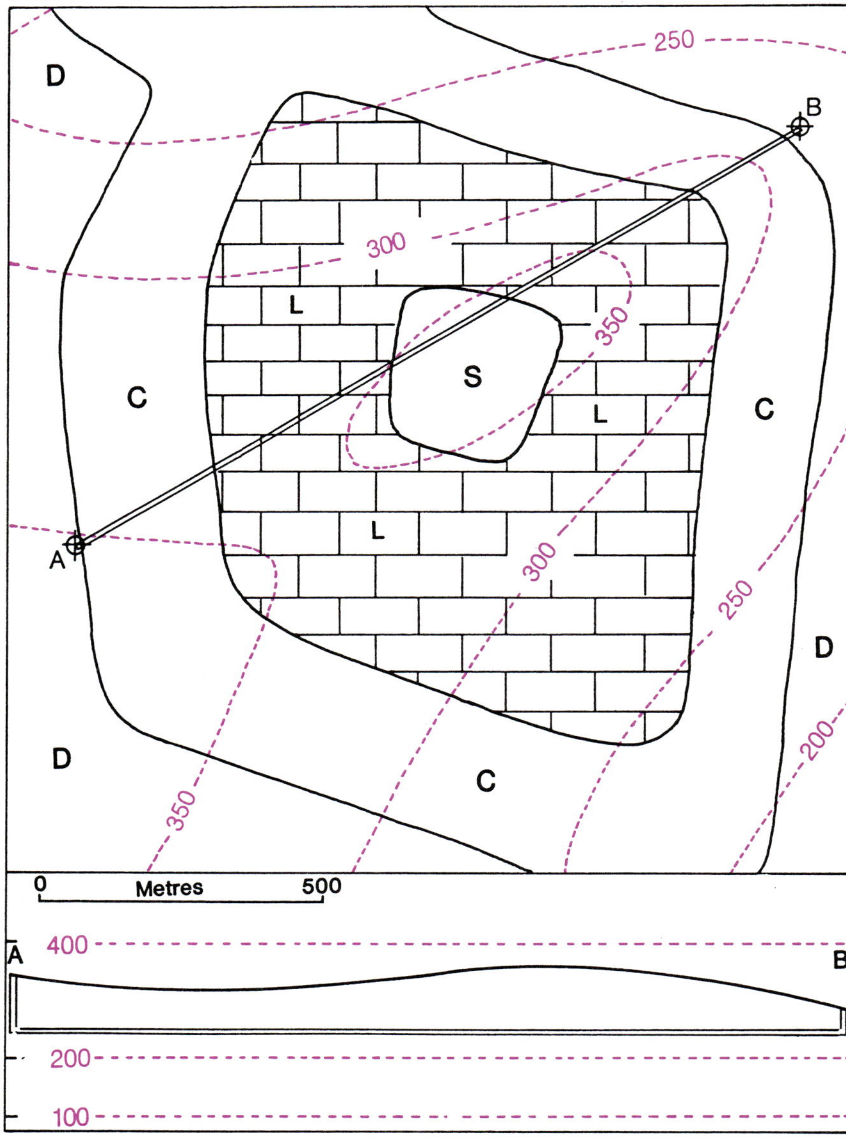

Problem Map A.3 Simple fold in tunnel.
Problem: Draw section AB to find whether the cavernous limestone (L) will occur in the tunnel, and if so, over what length. (Note: Invert is at 245 m above sea level.)

Problem Map A.4

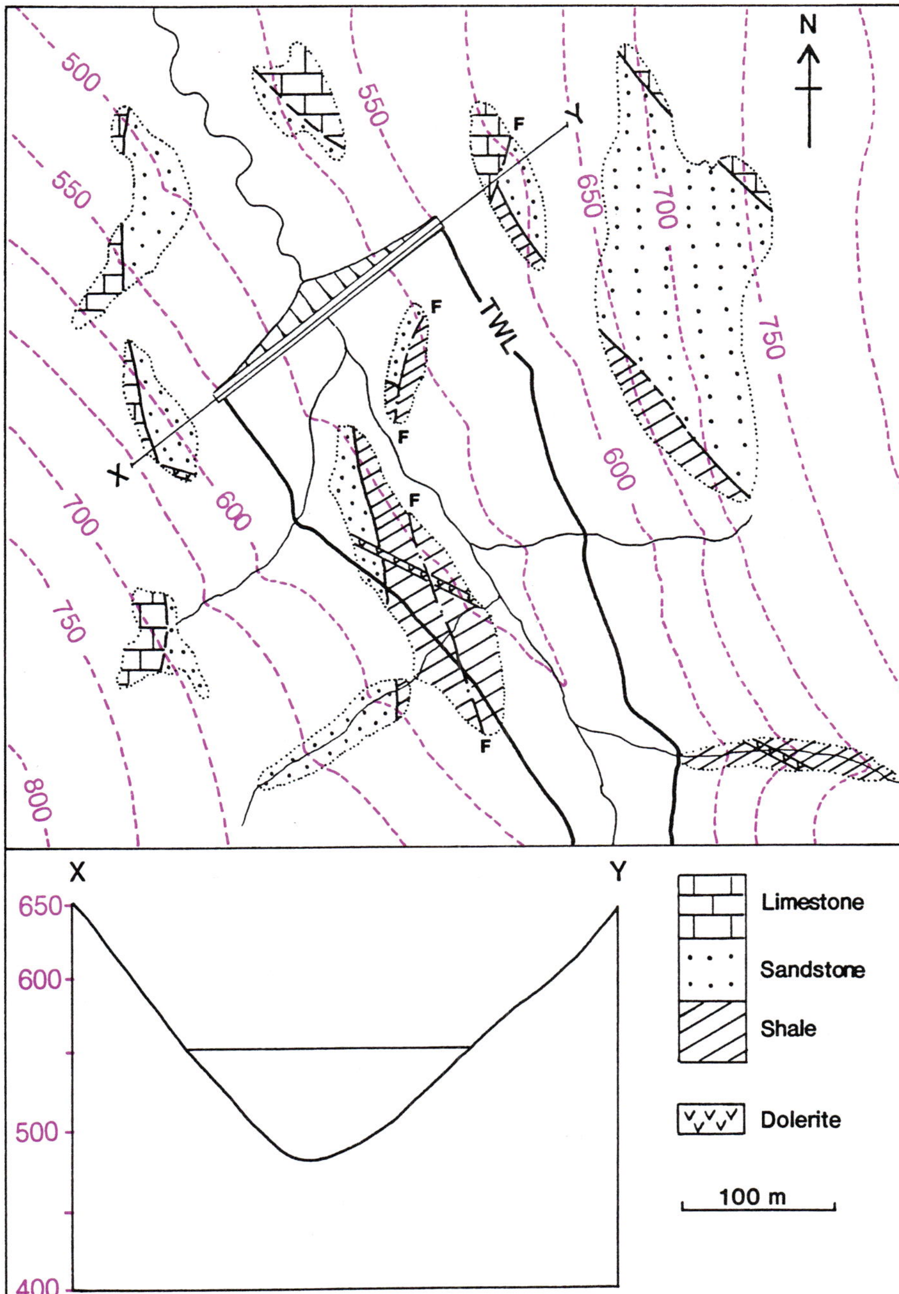

Problem Map A.4 Outcrops at a dam site.

Problem: (i) Complete outcrops from data on the map. (ii) Calculate the dip of the bedding and the fault, and the throw on the fault. (iii) Construct a geological section along the dam from X to Y, and comment on the suitability of the site.

Problem Map A.5

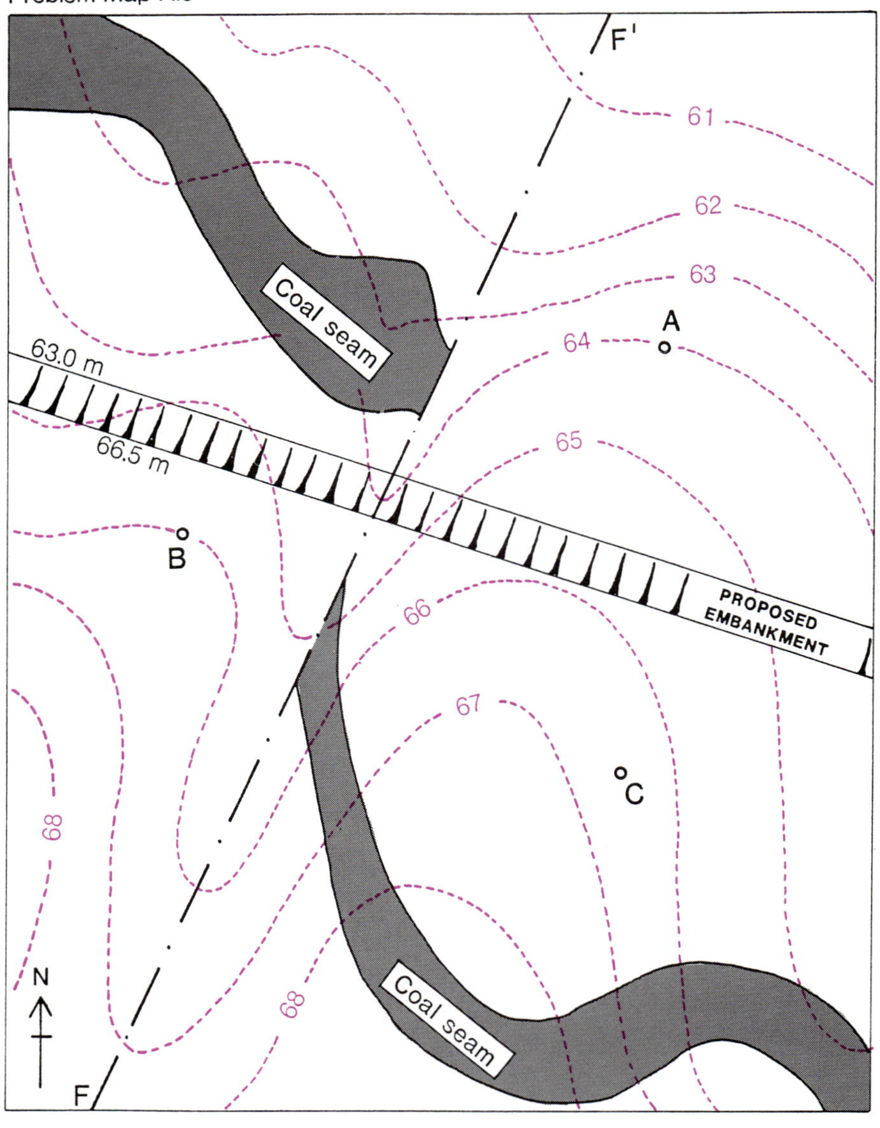

Problem Map A.5 Site benching.
A proposed factory site lying on a hill slope is to be benched flat at two levels, 63.0 m OD and 66.5 m OD, as indicated by the proposed embankment. A faulted coal seam 1 m in thickness has been proved to outcrop as shown.
Problem: (i) Calculate the angle and direction of dip, assuming it remains constant across the site. (ii) Plot the expected exposure of the coal seam after completion of the earth-moving operations on the benches. Shade in areas where fill will be required. (iii) Calculate the depth of coal under the benched site at points A, B and C. (iv) With reference to section 2.3.2 in Chapter 2, work out the apparent throw on the fault FF′.

Problem Map A.6

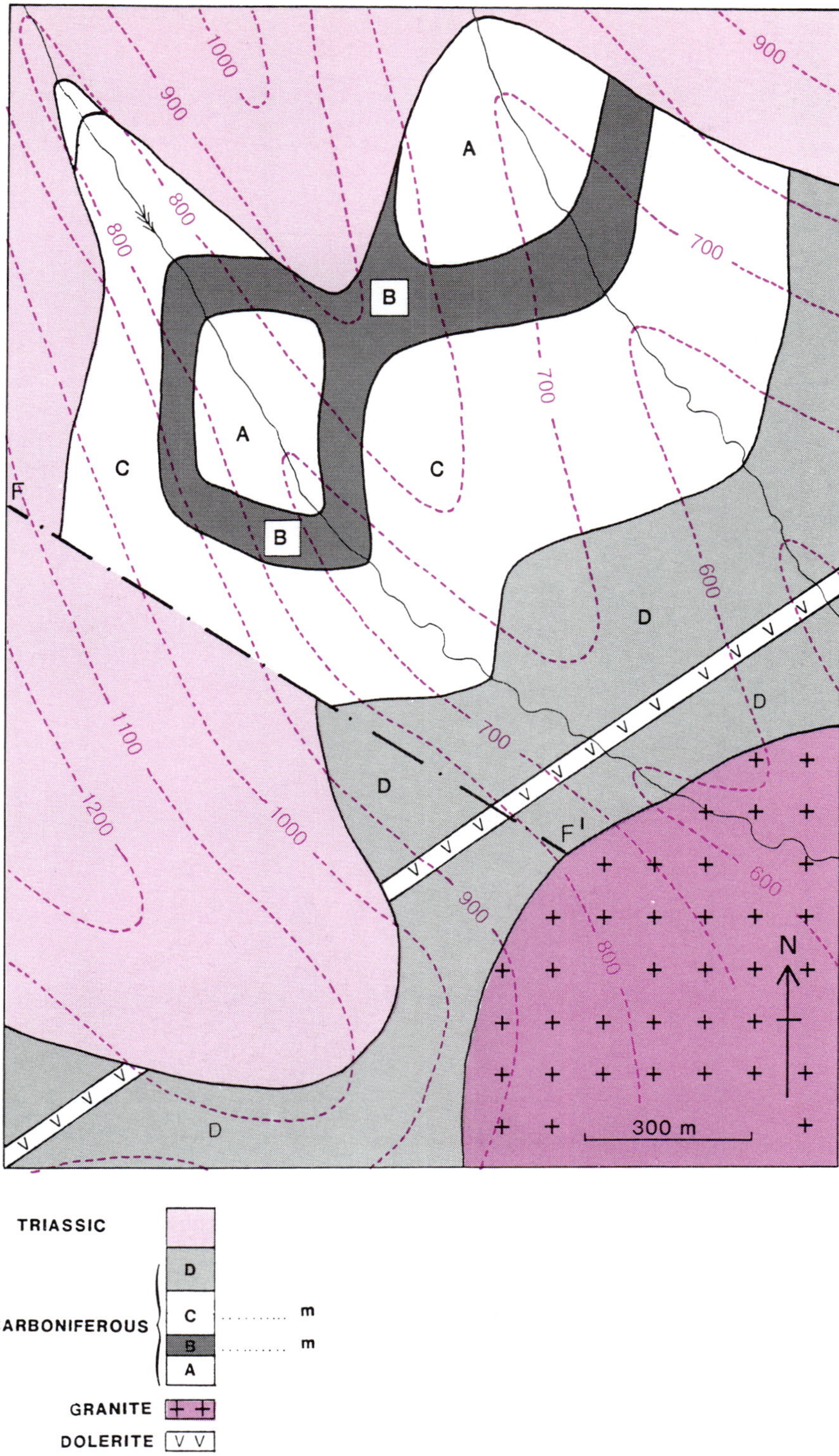

Problem Map A.6 General interpretation.
Problem: (i) Mark the surface of unconformity and the fold axis. (ii) Calculate the dips in the Triassic and Carboniferous strata. (iii) Calculate the stratigraphic thickness of B and C. (iv) Calculate the throw on the fault FF′.

Problem Map A.7

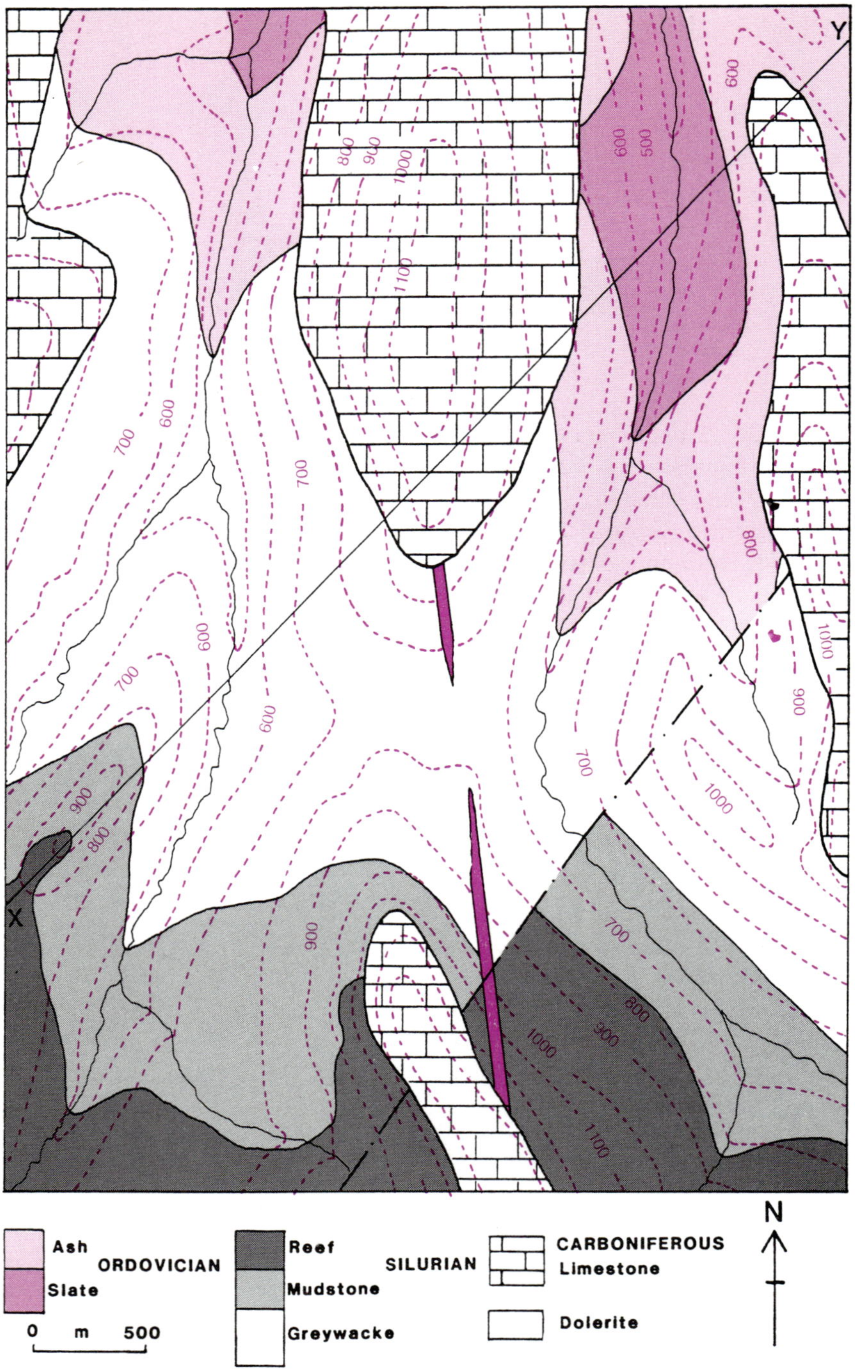

Problem Map A.7 Map analysis.
Problem: Write a geological report on the area, including the section XY. This map includes most of the elements discussed in Chapters 2 and 3.

Appendix B Engineering map symbols

The 1972 Working Party report of the Geological Society of London[4] suggested a comprehensive range of symbols for use on engineering geological maps. Many of their lithological symbols were adopted for BS 5930:1981 (Site Investigations)[12] with some notable differences to those recommended by the International Association of Engineering Geology in 1981.[21] (See Figure B.1).

(1) The symbol for silt (vertical stripes) adopted by the Americans and Scandinavians and recommended by the Geological Society, was replaced by an × symbol, since many rocks and soils contain subsidiary silt particles which made the vertical stripe more difficult to use.
(2) The IAEG peat symbol (black offset rectangles) was not adopted. Instead the generic plant symbol of the Geological Society was used.
(3) The IAEG sedimentary rock symbols may be confused with soils or certain other rocks (e.g. gypsum and basalt); BS 5930 reduces this confusion.
(4) Each group produced different igneous rock symbols, which has led to some confusion.
(5) The IAEG metamorphic symbols are really confusing whilst, once again, the WP and BS are too simple, not allowing for quartz content, marble, hornfels etc., but depending mainly on grain size.

Geomorphological and geodynamical features

There seems to be general agreement on geomorphological symbols. The British have not adopted so many geodynamical symbols, apart from landslides and solifluction sheets and these appear to be different.

Hydrogeological symbols

The Geological Society concentrates on defining formations as aquifers, aquitards, aquicludes and aquifuges (see Figure B.2). They define symbols for springs and sinks in a similar way to IAEG, but the latter body goes much further in differentiating between numerous spring types and also features water corrosiveness quadrants for pH, Cl, SO_4 and CO_2 values (see Figure B.2), used mainly in Czechoslovakia.

Structural symbols

The British obviously believe in detailed structural symbols for rocks since they appear on six pages of the Geological Society report. However, the succeeding BS gives only a selection of sixteen of these, but it is made clear that all arrows should be reserved for linear and fold structures, and planar structures should preferably be shown using a strike symbol. IAEG have not recommended any structural symbols on their engineering geological maps and this appears to be a major omission of both their 1976 and 1981 publications.[6,21]

Site investigation symbols

A comprehensive range of symbols for different types of borehole was suggested by the 1972 working party and although many of these have been used on site plans, there is some difficulty of application, since when both boring and drilling are carried out in the same borehole, the symbols can cancel each other out. However, because the basic concept is very useful, a revised set of symbols is suggested in Figure B.3. The geophysical traverse symbols and some in-situ methods can be differentiated on plans but it is difficult to adopt the sampling symbols, which may only show up on sections.

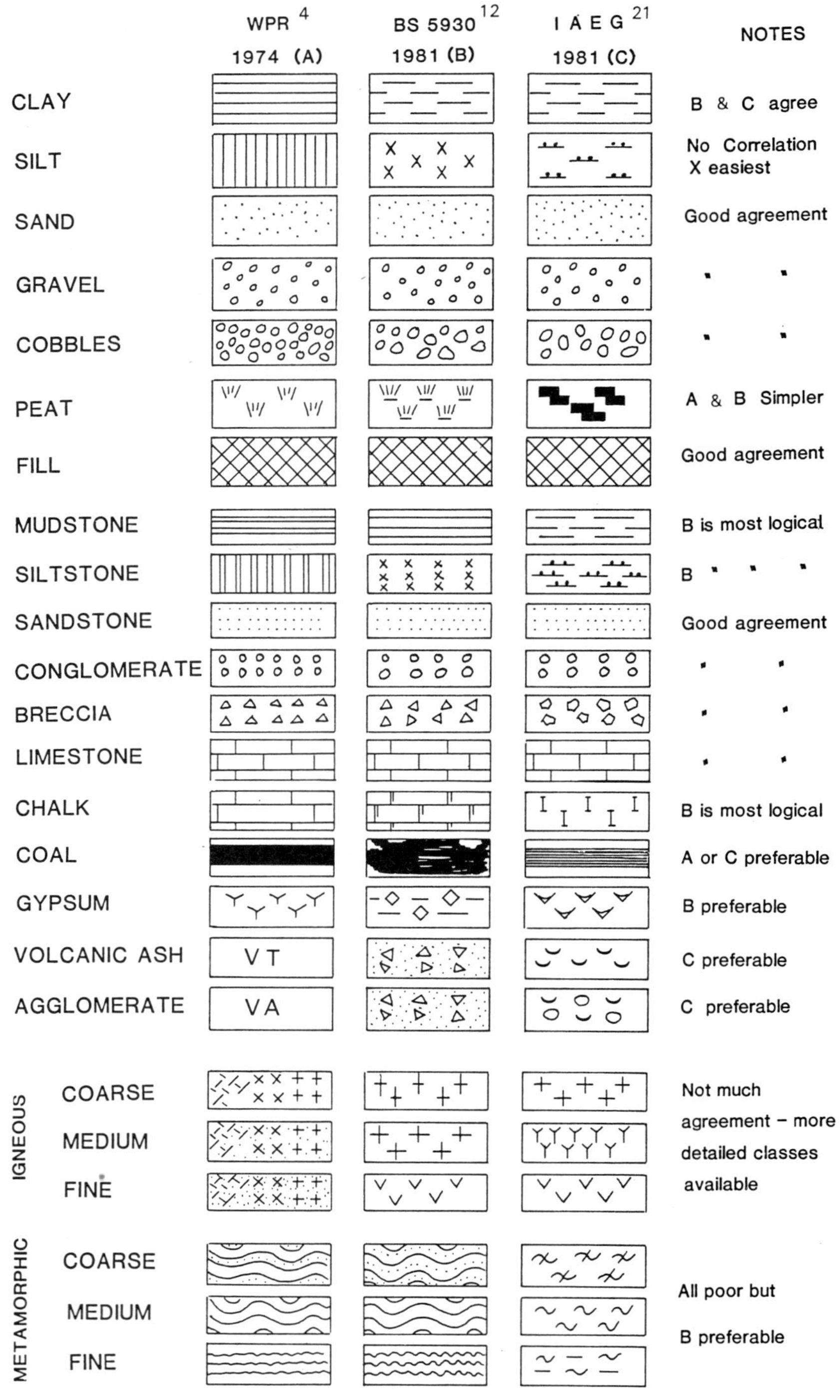

Figure B.1 Main lithological types.

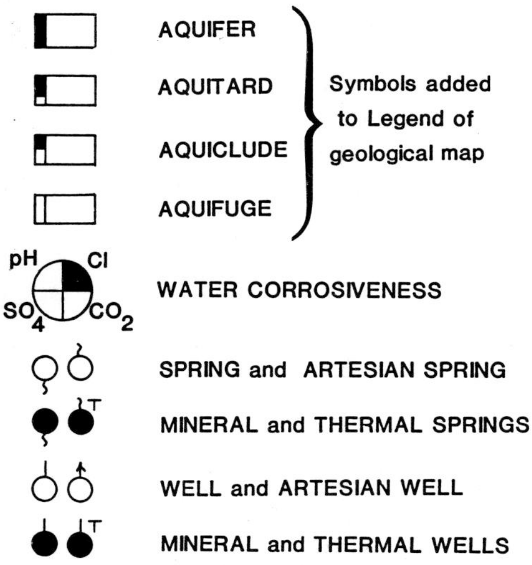

Figure B.2 Hydrogeological symbols. These would usually be represented with blue shading.

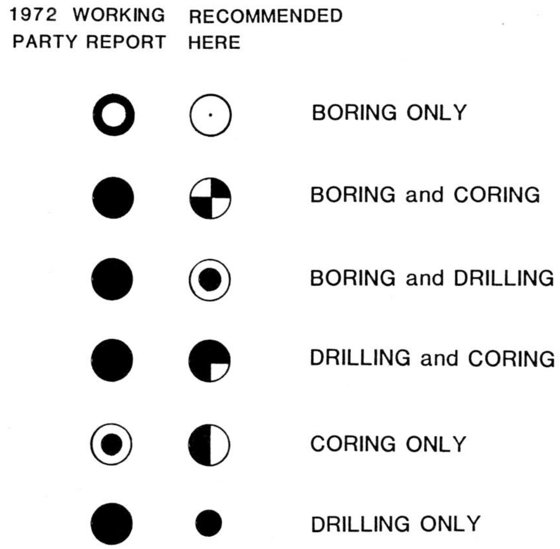

Figure B.3 Symbols for boreholes on plans.

Appendix C Stereographic projections

This method is used in conjunction with maps and sections to plot three-dimensional planar and linear structures in two-dimensions. From the stereographic plots some geometrical problems can be solved, statistical data recovered and displayed, and simple geotechnical analysis of rock stability attempted.

The projection is derived from an imaginary sphere, the lower hemisphere of which is used in structural and engineering geology. The projection sphere has a north pole to which all orientations are projected, so that they intersect and are shown on the equatorial plane of projection (Figure C.1).

Planes are represented by great circle curves (*cyclographic plots*) and lines are given as points.

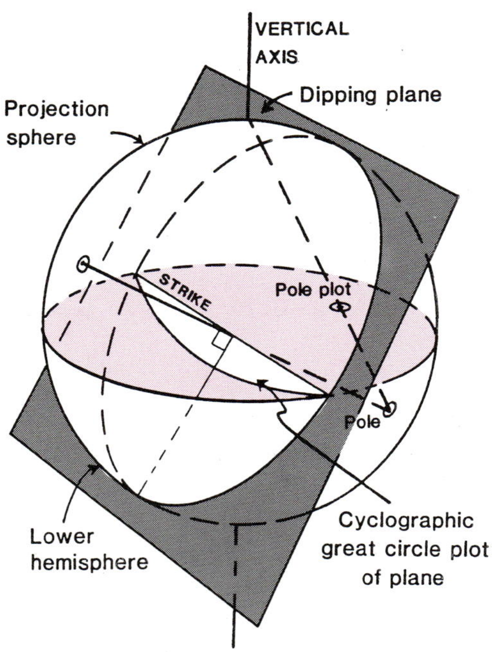

Figure C.1 The principle of the lower hemisphere stereographic plot.

It is simpler and less congested to use *poles* (or lines perpendicular to planes) if numerous discontinuities have to be added to the plot. The cyclographic plot of a vertical plane is a straight line through the centre whilst its pole plots would lie on the circumference.

An equal angle (*Wulff*) net is used to plot the geometry of lines and planes accurately. It has circular small circles marked every 2°. The Schmidt or equal area net (Figure C.2) has fourth-order quadratic curves for small circles which are particularly suitable for statistical work, where concentrations of poles of discontinuities can be contoured (Figure C.7).

Much of the statistical plotting can be carried out conveniently using a suitable computer package, but hand-plotting may be more convenient for solving simple geometrical problems using either net.

Plotting planes

For hand-plotting, it is normal to use a drawing pin upturned through the centre point and tracing paper as an overlay (but see TRRL Report No LR1039, Appendix D 13).

Outline the projection circle in pencil and mark N, S, E and W points. If a cyclographic plot of angle of dip α and direction of dip β is required, mark β° clockwise from N on the circumference and rotate the tracing so that the mark lies on the E–W line. Now mark in the strike as a straight line running N–S and count in from the circumference along the E–W line, the dip angle α. Then draw the great circle through that point to join the strike line at the circumference. To mark the pole of the plane, count a further 90° along the E–W line, and place a cross or point on the line. Where practicable, label each plane as it is plotted. It is worth noting that a net can be enlarged or reduced without affecting the results.

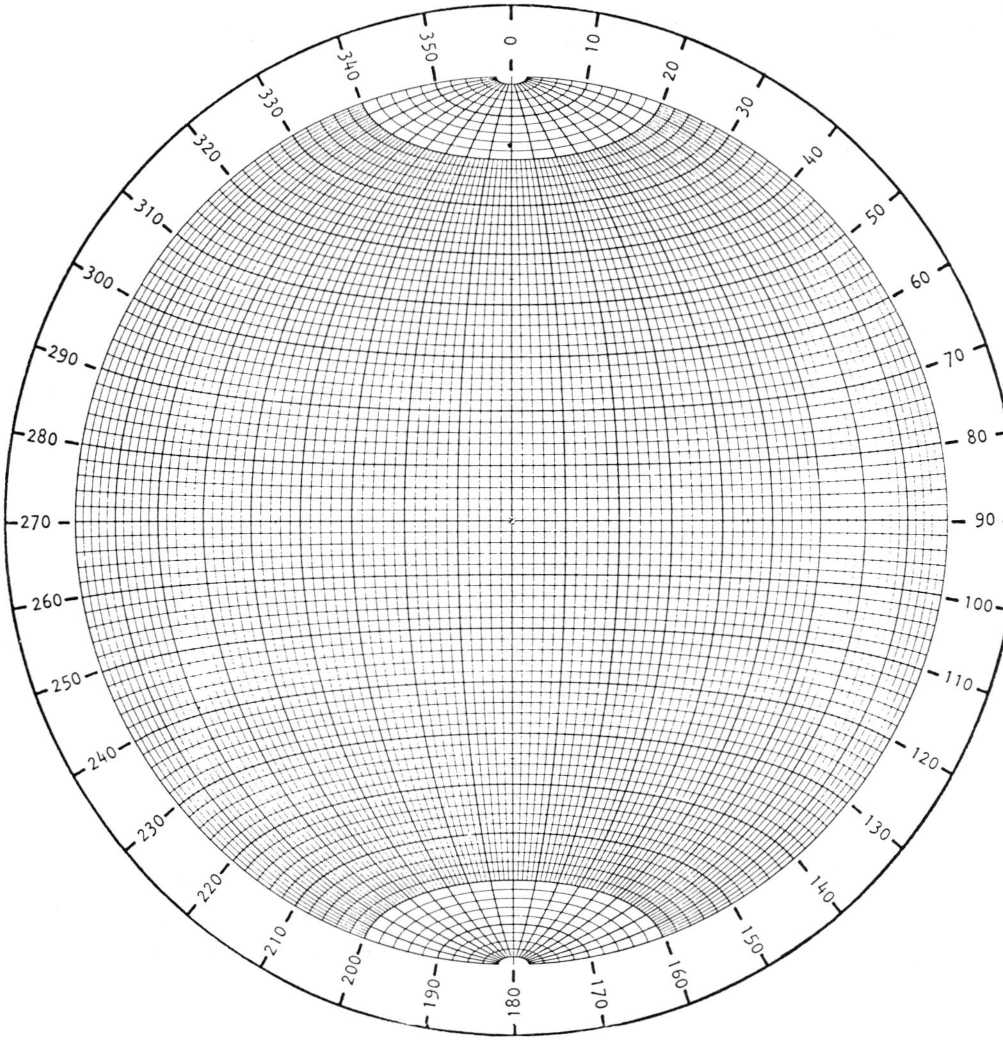

Figure C.2 Stereonet for use on problems in Appendix C. (Computer-drawn by C.M. St John of the Royal School of Mines, Imperial College, London.)

Examples of usage (use Figure C.2 and tracing paper)

1. *Apparent dip problems*

(a) Two adjacent faces in an excavation have apparent bedding dips of $\alpha_{A1} = 11/025$ and $\alpha_{A2} = 7/160$. Find the true dip of the bedding.

Solution: Mark the two apparent dips as linear (point) plots on tracing paper and rotate the net until a single great circle cuts both points. This is the cyclographic plot of true dip and should read as 22/086 (Figure C.3).

(b) Graphical method of finding the apparent dip (α_A) along the line of tunnel given true dip (α). If bedding on the map dips 48/330, find the apparent dip along an E–W tunnel line.

Solution: Draw the cyclographic great circle of the true dip. By rotating the tracing back to the N point, find the intersection of the

Example 1a

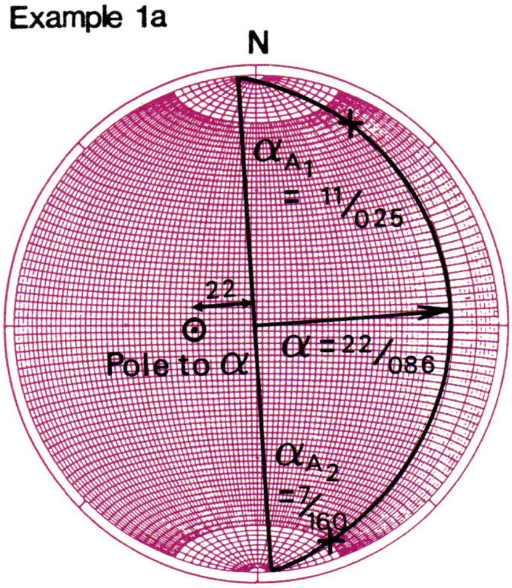

Figure C.3 True dip solution from direct measurement of two apparent dips. Note position of pole to true dip.

Example 2a

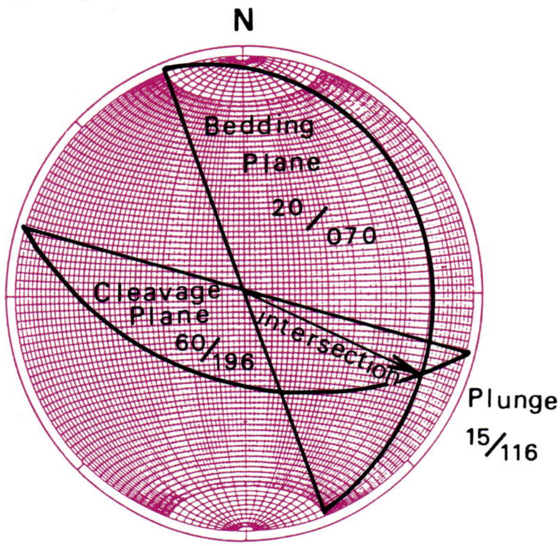

Figure C.5 Plunge of intersection of two planes.

great circle and the E–W line. Measure the dip from the circumference (29° to W) (Figure C.4). Equation (7) should give the same answer.

2. *Plunge of intersecting planes*

(a) *Fold plunge*: A slate exposure has a cleavage face dipping 60/196 with a bedding dip of 20/070. Find the plunge of the intersection of the two planes (which is also the direction of plunge of folds which may occur).

Solution: Plot both planes cyclographically and draw a straight line from the centre of the net to their point of intersection. This point represents the plunge angle (15°) and direction (116°) of the folds associated with the axial plane cleavage (Figure C.5).

Example 1b

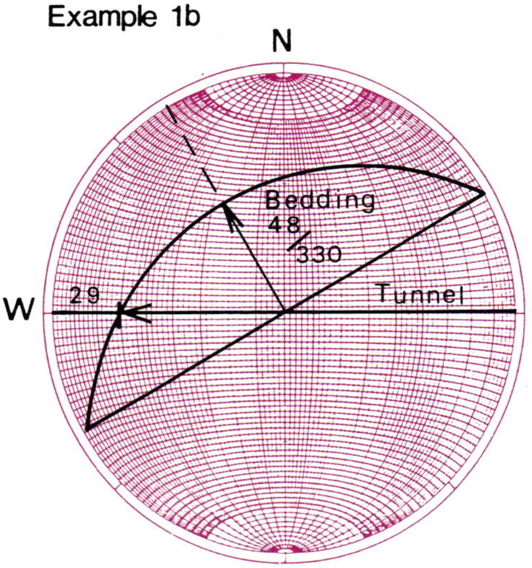

Figure C.4 Apparent dip along a tunnel line based on known orientation of true dip.

Example 2b

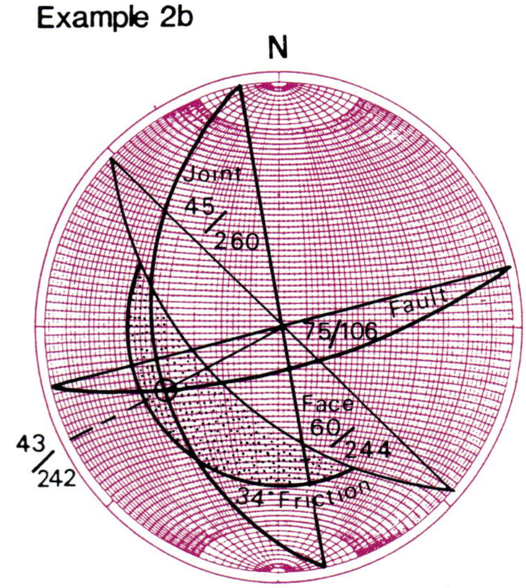

Figure C.6 Simple stability analysis for wedge failure in a rock face.

(b) *Wedge failure of a rock slope*: A rock slope of 60/224 contains a fault dipping 75/160 and a major joint dipping 45/260. Find the plunge of the wedge and assuming an angle of friction (ϕ) of 34°, state whether wedge failure is possible.

Solution: Plot the cyclographic curve of the slope, and add the friction circle 34° in from the circumference. The crescentic zone of sliding between the two has now been defined. If the intersection point between the fault and the joint lies within this zone sliding is possible. N.B. This is only a first approximation and more sophisticated analyses are available.[22] At 43°/242° the plunge of the wedge does lie within the zone of sliding (Figure C.6).

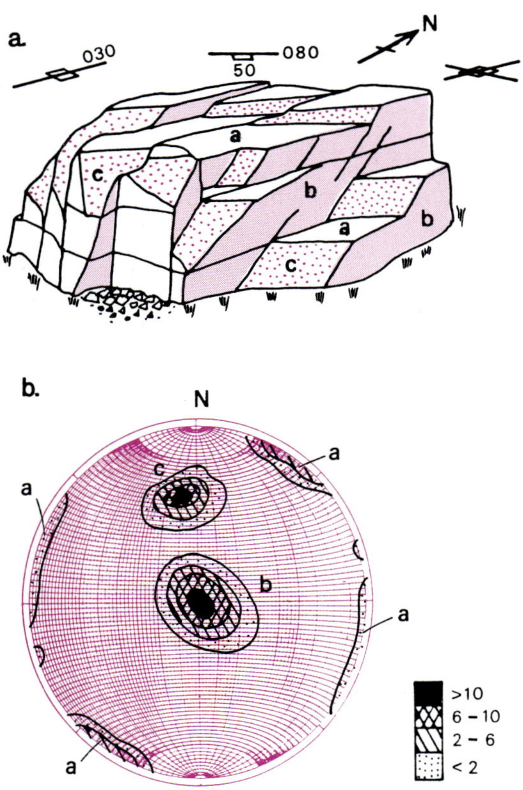

Figure C.7 Stereographic representation of joint sets in a rock mass. (*a*) Rock mass with three main joint sets: a = sub-horizontal, b = subvertical strike 030 (dominant open fractures), c = 40–60°/160–175°. (*b*) Contoured equal-area projection of the joint sets in (*a*). Originally plotted as poles.

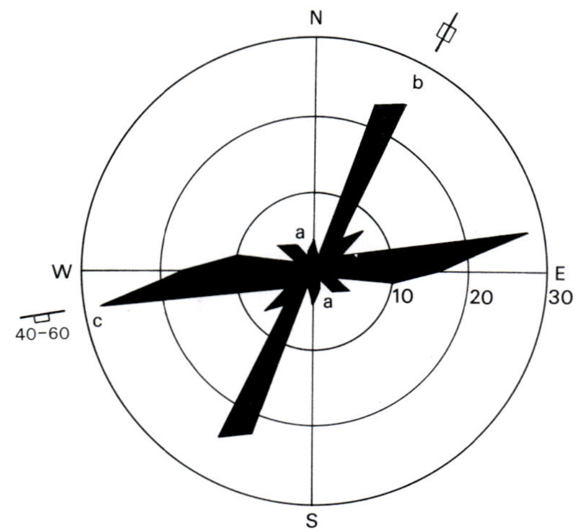

Figure C.8 Joint Rose Diagram for rock mass shown in Figure C.7(*a*).

3. *Discontinuity plots*

(a) *Statistical plots* of discontinuities measured in a rock domain (or area of similar structure) can be plotted as poles and contoured either by hand or more conveniently using a computer package (Figure C.7*a*, *b*).

(b) *Joint rose diagrams*: A circular net with radiating azimuths can be given unit lengths depending on the number of readings to be plotted. Only the strike of discontinuities can be shown; average dips of major sets have to be displayed at the circumference of the plot (Figure C.8).

Exercises

Solve the following problems using Figure C.2.

(1) Plot cyclographically and as poles the following bedding dips: 40/120, 62/233. What is the plunge of their line of intersection? (Answer: 32/162. Try to find this answer using the poles only.)

(2) Two apparent dips 22/183 and 26/246 are measured on rock faces. What is the true dip? (Answer: 28/221).

(3) From a map the true dip of bedding is 32/168. What is the apparent dip along a geological section trending 050°? (Answer: 17° in direction 230°).

(4) Two joints dipping 25/320 and 36/045 cut a rock face sloping 75/356. If the angle of friction $\phi = 30°$, is wedge failure a potential hazard? (Answer: The intersecting line between the joints is plunging at only 21° almost directly out of the slope, but this is well below the angle of friction of 30°. However, if the joints were clay filled or groundwater pressure was significant, cohesion and friction may be reduced to present a potential hazard.)

It is recognised that the above introduction is by no means exhaustive, but should be sufficient to give the engineer some understanding of the potential of stereonets. For more complex manipulations and rotations, specialist texts[22–24] should be consulted.

Appendix D Sources of further information

(1) **BRITISH GEOLOGICAL SURVEY**:
Maps, Memoirs and Reports

Main offices:
England: Keyworth, Nottingham, NG12 5GG
Tel. 06077 6111 Telex 378173 BGSKEYG
Fax 06077 6602
Scotland: Murchison House, Edinburgh, EH9 3LA
Tel. 031 667 1000 Telex 727 343 SEISEDG
Fax 031 668 2683
N. Ireland: Geological Survey of Northern Ireland, 20 College Gardens, Belfast, BT9 6BS
Tel. 0232 666595
London: Geological Museum, South Kensington, London, SW7 2DE
Tel. 01 589 4090

(2) **ORDNANCE SURVEY**

Topographical maps: Romsey Road, Maybush, Southampton, SO9 4DH

(3) **BRITISH STANDARDS INSTITUTION**

2 Park Street, London, W1A 2BS
BS 5930:1981 (Site Investigations), Cost £25

(4) **TRANSPORT AND ROAD RESEARCH LABORATORY**

England: Crowthorne, Berkshire, RG11 6AY
Scotland: Livingstone, West Lothian, EH54
For selected reports see later

(5) **SOIL SURVEYS**

England and Wales: Soil Survey and Land Resource Centre, Silsoe, Beds., HK45 4DT
Scotland: Macaulay Land Use Research Institute, Aberdeen, AB9 2QJ

(6) **AERIAL PHOTOGRAPHS**

Official

England: Air Photo Unit, Royal Commission for the Historical Monuments of England, Alexander House, 19 Fleming Way, Swindon, Wilts SN1 2NG. Tel: 0793 41400
Ministry of Defence f6t2 (Air)
St Georges House, St Georges Road, Harrogate, N. Yorks
Scotland: Central Registry of Air Photography of Scotland,
Scottish Development Department, Room 1/21, New St Andrew's House, St James Centre, Edinburgh EH1 3SZ. Tel. 031-244 4263
Wales: Central Register of Air Photography of Wales,
Welsh Office, Room G-003, Crown Office, Cathay's Park, Cardiff CF1 3NQ. Tel. 0222-823815
Overseas: Air Photo Library, Overseas Survey Directorate,
Room N.204. Ordnance Survey, Romsey Road, Southampton, SO9 4DH. Tel. 0703 792237

Commercial

BKS Surveys Ltd., 47 Ballycairn Road, Coleraine, Northern Ireland, BT51 3HZ
Cambridge University Collection of Air Photographs, The Mond Building, Free School Lane, Cambridge CB2 3RF
Clyde Survey Group, Clyde House, Reform Road, Maidenhead, Berkshire, SL6 8BU
Fisher-Spence Associates, 10 Drumfield Road, Holme Mains, Inverness, IV2 4XQ
Hunting Aerofilms Ltd., Gate Studios, Station Road, Boreham Wood, Herts, WD6 1EJ
Sealand Aerial Photography, Goodwood Airfield, Chichester, West Sussex, PO18 0PH
West Air Photography, 40, Alexandra Parade, Weston-super-Mare, Avon, BS23 2NG

(7) SATELLITE IMAGERY

National Remote Sensing Centre, Royal Aircraft Establishment, Farnborough, Hampshire, GU14 6TD

Nigel Press Associates, Edenbridge, Kent TN8 6HS

(8) HYDROGEOLOGY

British Geological Survey, MacLean Building, Crowmarsh, Gifford, Wallingford, OX10 8MR

(9) HMSO

Publications Centre, PO Box 276, London, SW8 5DT

Main bookshops: 49, High Holburn, London, WC1V 6HB; 13a, Castle Street, Edinburgh, EH2 3AR

British Regional Geology—Guides to 20 areas in Britain (see Figure D.1). Very useful background information on geology of Britain. Good value for money! Year of publication and 1989 prices (in £) are given below.

1. Orkney and Shetland	1976	2.25
2. The Northern Highlands Guide	1989	6.00
3. The Tertiary Volcanic Districts	1987	3.75
4. The Grampian Highlands	1978	1.75
5. The Midland Valley of Scotland	1985	5.00
6. The South of Scotland	1971	1.50
7. Northern England	1978	2.50
8. The Pennines and adjacent areas	1978	2.50
9. Eastern England	1980	4.50
10. Central England	1975	2.50
11. The Welsh Borderland	1981	3.50
12. East Anglia and adjoining areas	1982	2.50
13. London and Thames Valley	1960	out of print
14. The Wealden District	1979	2.50
15. The Hampshire Basin	1982	4.50
16. Bristol and Gloucester District	1981	3.25
17. South West England	1985	5.00
18. North Wales	1987	4.25
19. South Wales	1982	2.50
20. Northern Ireland	1986	6.50

(10) OVERSEAS GEOLOGICAL SURVEYS

United States Geological Survey: 1. Map Distribution, Federal Center, Box 25286, Denver, Colorado 80225, USA
2. Office of Earthquakes Volcanoes & Engineering, National Center, Stop 905, Denver, Colorado 80225, USA

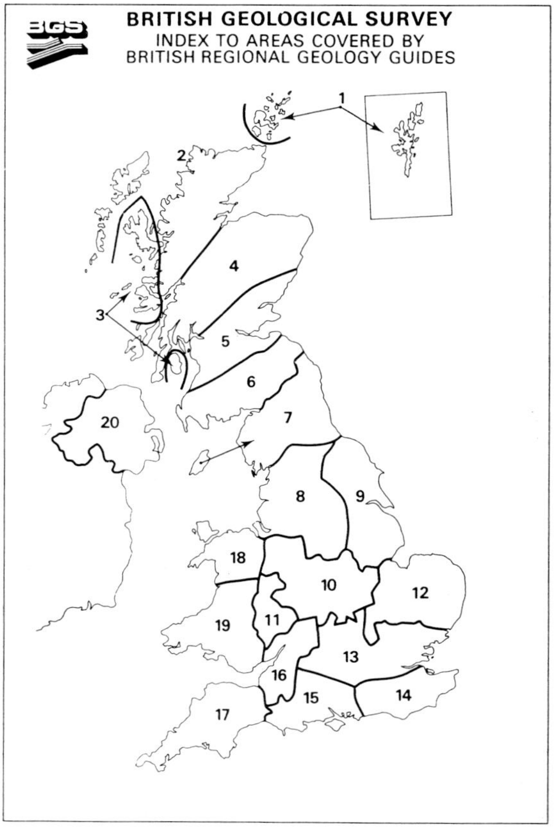

Figure D.1 Areas covered by British Regional Geology Guides, as listed in Appendix D. (Reproduced by permission of the Director, British Geological Survey. NERC copyright reserved.)

Geological Survey of Canada: 601, Booth Street, Ottawa, Ontario K1A OE8, Canada

Geological Survey of New South Wales: Department of Mineral Resources, P.O.B 5288, Sydney, NSW 2001

Geological Survey of Victoria: Department of Industry, Technology, and Resources, P.O.B.173, East Melbourne, Victoria 3002

Geological Survey of W. Australia: Mineral House, 100, Plain Street, Perth, Western Australia 6000

Geological Survey of S. Australia: Department of Mines, Adelaide

Geological Survey of India: 27, Jawaharlal Nehru Road, Calcutta 700013

Geological Survey of South Africa: Private Bag, X112, Pretoria, 0001

(11) ENGINEERING GEOLOGY TEXTS

Attewell, P.B. and Farmer, I.W. (1976) *Principles of Engineering Geology*. Chapman and Hall, New York.

Bell, F.G. (1980) *Geology and Geotechnics*. Butterworths, London.

Bell, F.G. (1983) *Fundamentals of Engineering Geology*. Butterworths, London.

Bell, F.G. (1975) *Site Investigations in Areas of Mining Subsidence*. Butterworths, London.

Bell, F.G. (Ed.) (1987) *Ground Engineers Reference Book*. Butterworths, London.

Blythe, F.G.H. and De Freitas, M.H. (1984) *A Geology for Engineers* (7th edn.). Arnold, London.

Brink, A.B.A., Partridge, T.C. and Williams, A.A.B. (1984) *Soil Survey for Engineering*. Oxford University Press, Oxford.

Clayton, C.R.I., Simons, N.E. and Matthews, M.C. (1982) *Site Investigation*. Granada.

ICE (1976) *Manual of Applied Geology for Engineers*.

Mathewson, C.C. (1981) *Engineering Geology*. Merrill, Ohio.

Weltman and Head (1982) *Site Investigation Manual*, CIRIA/PSA.

(12) ENGINEERING GEOLOGY JOURNALS

Quarterly Journal of Engineering Geology, London.

Bulletin of the International Association of Engineering Geology, IAEG

Engineering Geology, Elsevier.

Geotechnique, Institution of Civil Engineers.

Ground Engineering, British Geotechnical Society.

International Journal of Rock Mechanics and Mining Science, Pergamon.

Proceedings of the American Society of Civil Engineers, Geotechnical Division, American Society of Civil Engineers.

Geotechnical Abstracts, International Society for Soil Mechanics and Foundation Engineering.

(13) RELEVANT TRANSPORT AND ROAD RESEARCH LABORATORY (TRRL) REPORTS

LR 369 (1970) Air-photograph interpretation for road engineers in Britain, Dumbleton, M.J. and West, G.

LR 491 (1973) Available information for route planning and site investigation, Dumbleton, M.J.

LR 625 (1974) Guidance on planning, directing and reporting site investigations, Dumbleton, M.J. and West, G.

LR 702 (1981) Reproducibility of joint orientation measurements in rock, Ewan, V.J. and West, G.

LR 725 (1982) Terrain evaluation for highway planning and design, Beaven, P.J. and Lawrance, C.J.

LR 1039 (1983) Rock stability assessment in preliminary site investigations—graphical methods, Matheson, G.D.

Appendix E Outline answers to questions and problem maps in Appendix A

Question (1) The geological problems associated with the proposed dam site include:

(a) Strength, structure, and permeability of sandstones just below TWL at the northern abutment and in the centre of the valley (above unconformity), which could involve leakage problems.
(b) Steep tuff bands may be permeable.
(c) Southern half of the dam is on Coal Measures but no seams are indicated.
(d) Buried channel on south side of valley must be investigated for depth and content.
(e) Depth of boulder clay on south side unknown.

Question (2) Geological problems might include:

(a) Type of alluvial soils and peat in the valley and the existence of a buried channel.
(b) Coal seams (especially the K.Sp.) in a shaft at 30 m becoming deeper to the east, but shallower coals may also have been worked.
(c) Open cast fill in SW of site. Compaction and content must be checked.
(d) E–W fault through the middle of the site brings up the K.Sp. coal to the north, so that the north part of the site may not be suitable for any heavy loadings.
(e) Southern part of the site may be most suitable for development, but will require detailed site investigation.

Question (3) Problems which should have been considered at the site of the proposed tunnel include:

(a) Whilst the tunnel should be in rock most of the way down from the shaft, a piped section under the railway into bedrock may be necessary.
(b) Much of the tunnel down to the large fault near Dempster Burn will be in coal-bearing strata, with mines near the area of the shaft.
(c) The Millstone Grit (d^M) above the unconformity (approx. 1400 m from the shaft) may be easy to excavate, but could involve the ingress of large volumes of water into the tunnel.
(d) Below the unconformity stronger, steeply bedded greywackes, tuffs and slates striking ENE–WSW obliquely across the tunnel may involve some overbreak and underbreak problems, depending on the discontinuity pattern.

Question (4) Classes are (i) = 4, (ii) = 5, (iii) = 4, (iv) = 2, (v) = 1, (vi) = 3, (vii) = 1, (viii) = 3.

Question (5) Depth of soils are not available, but soil types and boundaries can be estimated and the degree of rock exposure determined. At sites of bridges and cuttings, bedrock becomes an important factor which should be highlighted for determination at the site investigation stage.

Question (6) Drift editions: New maps show marked improvement in clarity, which facilitates easier reading. No rock types are given so that a Solid edition must be consulted.
Solid editions: Colour printed on an improved base map. More detailed geology and probably more accurate boundaries.

Question (7) Problems include: unknown discontinuity data and mass permeability of the Red River Sandstone; position and character of the predicted fault zone; strength and structure of the greywacke; groundwater ingress at the unconformity; discontinuity patterns and boundaries of the granite; unknown rockhead and unconformity levels below the talus slope; roof and portal stability at both ends of the tunnel.

Question (8) Combine the data from all four extracts and describe using the geotechnical planning boundaries in (d).

Question (9) Whilst ease of excavation appears to be very favourable at the site of the dam, this may also indicate that rock mass strength may not be sufficient to support this high dam.

Map problems: numerical answers

Problem Map A.1a
Simple dipping strata dips 27/137. Section is not true scale, therefore use structure contours to plot 800 m quartzite.

Problem Map A.1b
Dip 14/237
Vertical thickness: limestone = 100 m
sandstone = 150 m
Length from B: Shale = 0–288 m
Limestone = 288–725 m
Sandstone = 725–925 m
Depth of siltstone/sandstone boundary in shaft A: 25
Apparent dip along the tunnel = 13 westerly.

Problem Map A.2a
Dip from 3 pt. solution = 7/172
Depth of the coal at pt. D = 1.25 m
Coal does not underlie the NNW part of the site except for a small outlier.
It is easier and just as instructive to construct an exaggerated section.

Problem Map A.2b
Dip = 14/120
The outcrop would allow a hydraulic connection in sandstone with the Lake Beds at 1500 m, 300 m below the proposed top water level. This could result in serious leakage depending on the mass permeability of the sandstone.

Problem Map A.3
Should cut into the limestone for about 190 m, from chainage 650 m to chainage 840 m. Exact position difficult to assess; use boreholes.

Problem Map A.4
Dip of bedding = 30/347
Dip of fault = 59/270
Downthrow on fault = 170 m to the west.
Geological problems are associated with the normal fault under the NE abutment, reservoir leakage and dam toe stability.

Problem Map A.5
Dip = 3/036
Depth at A = 6.33 m
B = no coal
C = 2.5 m
Once benched, coal outcrops should be straight, but terminate at the fill boundary.
Fault downthrow = 3.3 m to the SE.

Problem Map A.6
Dip of Carboniferous = 45/130 or 45/310
Dip of Triassic = 22/360
Anticline NE–SW in the valley is cut by low dipping unconformity.
Thickness of B = 100 m
C = 300 m
Later fault FF′ downthrows 347 m to the SW.
The dyke is earlier than the unconformity, and is cut by the fault without any apparent displacement. The granite pluton is even later as it cuts the fault.

Problem Map A.7
Dip of Ordovician–Silurian rocks is 23/226 or 23/046.
Dip of the Carboniferous limestone is 8/360.
The Carboniferous is unconformable on the older rocks, which are gently folded on a horizontal NW–SE axis.
The thickness of the mudstone = 300 m, the ash = 400 m, and the greywacke = 950 m (using the dip method).
A NE–SW trending vertical fault downthrows the older strata 650 m to the SE, but is older than the Carboniferous.
The section should clearly show the older folded rocks, gently dipping unconformity on two hills and the projected position of the vertical dolerite dyke.

Index